BORDEN INSTITUTE MONOGRAPH SERIES

Musculoskeletal Injuries in Military Women

Colonel Barbara A. Springer, PhD, PT, OCS, SCS,
and Major Amy E. Ross, MD

EDITED BY
Paul J. Dougherty, MD
Colonel, MC, US Army Reserve

BORDEN INSTITUTE

Musculoskeletal injuries in military women are common. Prevention and management of such injuries are very important to sustain the fighting force and maintain military readiness. This monograph provides information about the incidence, risk factors, prevention, diagnosis, evaluation, treatment, and rehabilitation of common musculoskeletal overuse and traumatic injuries sustained by women in the military.

Published by the Office of The Surgeon General
Borden Institute
Fort Detrick, Maryland
US Army Medical Department Center & School

CONTENTS

FOREWORD v

ACKNOWLEDGMENT vii

INTRODUCTION 1

EPIDEMIOLOGY 3
- Overview of Musculoskeletal Injuries in Military Women 3
- Common Lower Extremity Overuse Injuries in Military Women 5
  - Bone 5
  - Tendon 6
- Common Traumatic Injuries in Military Women 22

PREVENTION, TREATMENT, AND REHABILITATION 37
- Overview of General Injury Prevention, Treatment, and Rehabilitation Techniques in the Military 37
  - Extrinsic Risk Factors 37
  - Intrinsic Risk Factors 37
  - General Prevention Strategies 38
  - Running Shoes 39
  - Multiple Intervention Programs 41
  - General Treatment and Rehabilitation Strategies 44

Specific Injury Prevention, Treatment, and Rehabilitation Techniques in the Military 46

SUMMARY 91

REFERENCES 93

APPENDIX: POSTOPERATIVE PROTOCOL—HIP ARTHROSCOPY FOR ACETABULAR LABRAL LESIONS 107

Cover photo by Staff Sergeant Amanda Smolinski, Fort Bragg, North Carolina

Barbara A. Springer is a Colonel, Army Medical Specialist Corps, US Army; Director, Proponency Office for Rehabilitation & Reintegration, Office of The Surgeon General, Falls Church, Virginia 22041; formerly, Chief, Integrated Physical Therapy Service, National Naval Medical Center, 8901 Rockville Pike, Bethesda, Maryland, and Walter Reed Army Medical Center, 6900 Georgia Avenue, NW, Washington, DC 20307. Since publication of this monograph, Colonel Springer has retired.

Amy E. Ross is a Major, Medical Corps, US Army; Orthopaedic Surgeon, 121 Combat Support Hospital, Seoul, Korea, Unit 15244, Box 338, APO, AP 96205. Since publication of this monograph, Major Ross has been promoted to Lieutenant Colonel.

Foreword

The percentage of women and roles these women occupy in the military have greatly expanded in the last 10 years. Department of Defense data from September 2010 indicate that women comprised 14.5% of our active duty service members and 17.7% of the Reserve and National Guard service members. Whether these women are serving during a conflict or not, they are employed in a wide range of military occupations that may put them at risk for injuries. This comprehensive monograph by Drs Springer and Ross addresses the incidence, risk factors, prevention, diagnosis, evaluation, treatment, and rehabilitation of common musculoskeletal overuse and traumatic injuries sustained by women in the military.

The authors emphasize that, anatomically and physiologically, women are not the same as men; lower extremity biomechanical differences between men and women may account for gender differences in training injury rates. Women have increased pelvic width, forefoot pronation, heel valgus angulation, pes planus, external tibial torsion, and femoral anteversion. Additionally, because of the estrogen influence, women have less lean body mass and greater ligamentous laxity. The combination of anatomy and physiology appears to predispose women to a higher risk of pelvic stress fracture and anterior cruciate ligament (ACL) tears. The diagnosis of pelvic stress fracture has been reported as 1 in 367 female recruits, compared with 1 in 40,000 male recruits, and rates of ACL ruptures for female athletes range from 2.4 to 9.7 times higher than in male athletes.

This monograph is an excellent reference for lower extremity and back injuries, and contributes significantly to understanding the differences between women and men in the rate and distribution of musculoskeletal overuse and traumatic injuries sustained by our military members. This publication sheds light on the nature of the biomechanical differences between women and men, and how these differences result in notable changes in injury patterns that impact the length of healing time and ultimately impact the readiness of the unit. Drs Springer and Ross have provided us with an excellent tool for addressing the unique challenges healthcare providers face in providing the best possible care—both therapeutic and preventive—to our women military members.

Sonja M. Thompson, MD, Colonel, MC, US Army
Chief of Surgery, Carl R. Darnall Army Medical Center
Fort Hood, Texas

Acknowledgment

We thank Major Norman (Skip) Gill for his assistance with reviewing and editing this manuscript.

Introduction

The role of women in the military has increased over the years. A number of legislative changes led to the expanded role for women in today's military. The first major factor was a shift in the mid-1970s from the military draft to an all-volunteer force, which created opportunities for women to serve in greater numbers as the services sought to meet personnel goals. In the early 1990s, Congress lifted the ban on women flying combat aircraft and serving on combat ships, and during the first Clinton administration, then Secretary of Defense Les Aspin announced new rules and policies that opened more military jobs to women. Even though women are prohibited from serving in direct combat, they are finding themselves in the thick of battle because of the unpredictable nature of recent conflicts and the blurred distinction between frontline and rear areas. Whether they are serving during a conflict or not, women are employed in a wide range of military occupations that may put them at risk for injuries. Musculoskeletal injuries can be very detrimental to military women, especially if they lead to lost work time, outpatient visits, hospitalization, functional limitations, and/or disabilities.

Epidemiology

Overview of Musculoskeletal Injuries in Military Women

Army women are more likely to be disabled than men and are approximately 67% more likely than Army men to receive a physical disability discharge for a musculoskeletal disorder.[1] The discharge rates for musculoskeletal conditions have been as high as 140 per 10,000 Army women per year, compared with 81 per 10,000 Army men per year.[1]

Military women tend to suffer a higher incidence of injuries than military men. Several studies have identified female gender as a risk factor for injury in Army basic training programs in the United States and around the world.[2–10] For example, one study shows the cumulative injury incidence in Basic Combat Training (BCT) was 52% for women versus 26% for men. It was 30% for women versus 24% for men in Advanced Individual Training (AIT).[11] Other studies showed a similar incidence for training injuries in BCT populations: approximately 50% for women and 25% for men.[4,10,12] In addition, the proportion of trainees discharged from BCT for medical reasons was 12.7% for women, compared with only 5.2% for men.[13] There was even reported gender differences in the utilization of medical services on a military ship.[14] During a 6-month period, females were evaluated at a rate 9.2 times that of males (6.44 vs 0.70 visits per year). Only 39% of the visits were gender-specific, whereas gender-neutral conditions resulted in a female-to-male visit ratio of nearly 6:1.[14]

Lower extremity injuries are the most common in BCT and AIT, and account for 79% to 88% of new injuries for women.[4,15,16] Lower extremity biomechanical differences between men and women may account for gender differences in training injury rates. Compared with men, women have increased pelvic width, forefoot pronation, heel valgus angulation, pes planus, external tibial torsion, and femoral anteversion. Furthermore, women often have a quadriceps angle greater than 15 degrees and a hypoplastic vastus medialis obliquus.[17] There are also significant differences in physical performance after ages 10 to 12. Women reach skeletal and physiological maturity before males. They have more body fat and less lean body mass than males, which are attributed to increased estrogens in women and increased androgens in males.[18]

In the Army, sports and physical training are associated with the largest proportion of injuries in men and women.[1] Acute and chronic musculoskeletal problems associated with injuries are consistently the leading causes of outpatient visits and hospitalizations in the Army, and the Army's rate of such types of hospitalizations is more than twice that of the Air Force and almost three times that of the Navy.[1] The third leading cause of musculoskeletal injury requiring hospitalization in the Army is physical training, athletics, and sports combined.[1] On BCT installations, the incidence of injury is 1.4 to 2.2 times higher than the overall Army installation average.[1] Running, in particular, seems to be the primary physical activity associated with overuse injuries in BCT and AIT[19] and in Army occupations.[20] Other possible causes include marching, walking, drill and ceremony, and jumping.[21] Possible causes of injury in the military include the following:

- initial entry training (BCT and AIT);
- field training exercises;
- motor skills training;
- physical training
  - running,
  - calisthenics,
  - marching,
  - jumping;
- military training (airborne, air assault, etc);
- specific military occupational specialties; and
- sports, weight lifting, and recreation.

Common Lower Extremity Overuse Injuries in Military Women

Overuse injury is typically "due to or related to long-term energy exchanges resulting in cumulative microtrama."[20] Usually, it occurs as a result of repetitive overload forces on bones, ligaments, tendons, and muscles, which are unable to successfully adapt to the stresses. Examples include the following:

- stress reactions and fractures,
- shin pain,
- patellofemoral syndrome,
- patellar tendinitis or tendinosis,
- iliotibial band (ITB) friction syndrome (ITBFS),
- Achilles tendinitis or tendinosis, and
- low back pain.

Military activities that may contribute to overuse injuries include running, marching (including road marching), and repetitive jumping. Approximately 75% of all injuries in BCT can be classified as overuse injuries.[6,16] Overuse injuries can also occur once a woman completes her initial military training and is working in her specialty. In a study of Army wheel vehicle mechanics, overuse injuries accounted for 68% of female injuries (compared with 48% of male injuries).[20] Bone and tendons are examples of tissues that are often involved in overuse injuries.

Bone

Bone tissue is a very dynamic substance and is constantly remodeling in response to mechanical changes in the environment (eg, muscular forces, tension, and compression) and nonmechanical factors (eg, preexisting physical activity level, drugs, hormones, nutrition, race, gender, smoking, disease, inflammation, and infection). Normally, the rate of bone resorption equals the rate of bone regeneration, and the bone is in a steady state. Stress fractures occur when there is an imbalance between bone resorption and bone regeneration caused by mechanical, hormonal, and nutritional factors.[22] Normally, bone will adapt to changes in load frequency and magnitude if the mechanical changes or stresses are increased gradually over time. However, if the process of a bone's adaptation to a new stress shifts out of balance (more resorption occurs prior to efficient addition of new bone), the possibility of microfracture or catastrophic failure increases.[23] Lower extremity stress fractures in military personnel are considered overuse injuries because they result from bone structural failure caused by repetitive weight-bearing activities, such as running, walking, and marching. The bone tissue is not able to regenerate adequately to withstand the magnitude of loading and/or the frequency associated with these activities.

Tendon

Tendinitis and tendinosis are also common overuse conditions in military women. The tendons that are most often vulnerable to tendinopathy (a collective term that includes both tendinitis and tendinosis) in BCT include rotator cuff, biceps brachii, patellar, quadriceps, and Achilles tendons.[21] Acute tendinitis is inflammation of a tendon with increased cellularity and vascular disruption. It primarily involves inflammation of the outer layer of the tendon: the paratenon. Acute tendinitis is characterized by acute swelling, pain, warmth, local tenderness, and minimal dysfunction. It occurs when repetitive stresses are applied to the tendon before the tendon is allowed to fully recover. If left untreated and the external repetitive forces continue, it can lead to chronic tendinitis, which is increased tendon degeneration and hypervascularity. It is characterized by chronic pain, local swelling and tenderness, and increased dysfunction (eg, decreased loading). If the microtrauma continues even further, chronic tendinitis can lead to tendinosis. Tendinosis is best described as intrasubstance degeneration of the tendon, vascular compromise, collagen disorganization and disruption, and focal necrosis. Clinical signs of tendinosis include palpable tendon enlargement, swelling of adjacent connective tissues, increasing levels of dysfunction, and swelling of the tendon sheath.[24]

Lower Extremity Stress Fractures

Incidence/Risk Factors. Stress fractures, also known as fatigue or march fractures, are defined as a partial or complete fracture of bone that results from the repeated application of a stress lower than that required to fracture the bone in a single loading situation.[25,26] Described by Armstrong et al[27] (but initially by Breithaupt[28] in 1855), it is a commonly diagnosed overuse injury in athletes and is frequently seen in the physically active military population. Although it is often suggested that women in general sustain a disproportionately higher number of stress fractures than men, the studies that directly compare stress fracture incidence in male and female athletes show either no difference or only a slightly increased risk for women.[25,29–34] This equitable distribution of injury incidence does not hold true for the military population, wherein the risk of stress fracture in female recruits ranges from 1.2 to 10 times that of men.[23,35–40] Possible anatomical and physiological risk factors include lower bone density; differences in gait; more slender bones with lower moments of inertia; unfavorable biomechanical conditions, including a wide pelvis, coxa vara, genu valgum, tibia torsion, increased hip external rotation, and hyperpronation of the subtalar joint; greater percentage of body fat loading the musculoskeletal system; endocrine factors relating to body somatotype, amenorrhea, and oligomenorrhea; and lower initial physical fitness.[25,41–44] Previous studies of military populations have identified low levels of aerobic fitness,[12,45,46] endocrine dysfunction,[46,47] and white race[48] as risk factors among women entering basic training.

Military-Specific Risk Factors. Military boots are designed to provide good support and minimize ankle sprains, but they lack adequate shock absorption when compared with running shoes.[49] Well-designed footwear may reduce the incidence of lower extremity stress fractures by reducing excessive shock loading by as much as 33%.[49] One prospective study demonstrated a reduction in the incidence of overuse injuries during military training by adding a neoprene shock-absorbing insole to the military combat boot.[50]

It is estimated that 5% to 12% of women undergoing various entry-level military training programs sustain a stress fracture.[51] The lower extremities are the most common sites of stress fractures and include the pelvis, femur, tibia, and metatarsals.[27,51] Pelvic and femoral stress fractures, in particular, are more common among women than men in the military. Femoral and pelvic stress fractures require a more prolonged rehabilitation period (approximately 4 months, compared with 1 to 2 months for tibial, fibular, or metatarsal stress fractures)[52,53] and have a higher frequency of complications, including delayed union, and non-union, and avascular necrosis.[54,55]

Most military recruits report the onset of symptoms of stress fractures early in the training cycle, frequently between days 10 and 12, which may represent structural fatigue produced from repetitive mechanical forces secondary to the new onset of training.[44] It has been hypothesized that as excessive, repetitive, submaximal mechanical forces are placed on bone, the rate of osteoblastic formation of bone is exceeded by osteoclastic resorption. This process continues until the stress ultimately exceeds the limit of the bone and fracture occurs.[44] Additionally, an abrupt increase in the duration, intensity, or frequency of physical activity without adequate periods of rest may lead to an escalation in osteoclastic activity and ultimate fracture.[56]

Pathophysiology. A progressive decline in the muscular support of the bone, secondary to muscle fatigue, may lead to the transmission of excessive forces to the underlying bone. Muscles that are not adapted to repetitive work, and therefore lack endurance and muscle mass,[14] may be unable to support the long bones of the lower extremity. Muscles may also contribute to stress injuries by concentrating forces across a localized area of bone, thus causing mechanical insults that exceed the stress-bearing capacity of the bone.[57]

The aforementioned pathophysiology of stress fractures is a simplified model; however, other physiological and anatomical factors, such as those mentioned previously and others that are beyond the scope of this text, ultimately contribute to the occurrence of a stress fracture. The endocrine system plays a vital role in bone health. Male and female competitive endurance athletes with abnormally low sex hormone levels are predisposed to stress fractures.[56] The "female athlete triad"—which refers to the combination of amenorrhea, osteoporosis, and disordered eating—may predispose a female to stress fractures. In attempts to minimize body fat to further increase athletic performance, a female may find herself in an estrogen-deficient state leading eventually to decreased bone mineral density and increased risk of stress fractures.[56] Although not specific to military women, amenorrhea and oligomenorrhea are common findings in competitive female distance runners, with the prevalence of menstrual irregularities as high as 50%.[56,58]

Diagnosis. The history of a patient with a stress fracture is typically one of insidious onset of activity-related pain. The pain is generally well localized and described as a mild ache occurring after exercise. As time and activity participation continue, the patient may report more severe pain or pain that occurs at an earlier stage in exercise. The discomfort may ultimately limit the quality or quantity of physical activity participation or may require cessation of all activity as pain increases in duration and intensity. A full history (including exercise routine, diet, and menstrual pattern) is important in understanding the patient's injury and risk factors.

The most obvious finding on physical examination is localized bony tenderness, which may also be accompanied by periosteal thickening, redness, and swelling if the stress fracture occurs in a superficial area of the body.[25,56] The physical examination should include evaluation of limb biomechanics to identify potential predisposing factors (eg, leg-length discrepancy or malalignment, muscle imbalance, weakness, excessive subtalar pronation, or lack of flexibility).[25,56] The differential diagnosis of stress fracture may include nonbony pathology, such as exertional compartment syndrome, nerve entrapment, muscle strain, bursitis, traction periostitis, or medial tibial stress syndrome. Bony pathologies that can mimic stress fracture include infection and neoplasm.

Although a classic history of exercise-associated bone pain and typical examination findings of localized bony tenderness have a high correlation with the diagnosis of stress fracture, various imaging techniques are also available to the clinician for further evaluation.[25] Additional diagnostic imaging studies include radiography (plain X-ray), bone scintigraphy (bone scan), computerized tomography, and magnetic resonance imaging (MRI).

Radiographs are typically normal for the first 2 to 3 weeks after the onset of symptoms and may not reveal positive findings, such as periosteal reaction, cortical lucency, or a fracture line for several months.[56] Therefore, radionuclide imaging (bone scan), which is highly sensitive for detecting stress injuries,[59] may be used to confirm a clinically suspected stress fracture. Acute stress fractures may be depicted as discrete, localized, linear areas of increased uptake involving one cortex or extending the width of the bone, which is seen on all three phases of a technetium-99m diphosphonate bone scan.[59,60] Changes may be seen as early as 48 to 72 hours after the beginning of symptoms. Other bony abnormalities, such as periostitis, are only positive on delayed images,[61,62] and other soft-tissue injuries may only be positive in the angiogram and blood pool phase.[25]

Bone scintigraphy is more sensitive than MRI and may be useful in evaluating suspected lesions in the spine and pelvis, identifying multiple stress fractures, and distinguishing bipartite bones from stress fractures.[56] Caution should be utilized with technetium bone scanning because, although this test is sensitive, it is not specific for stress fractures. Additionally, bone scans should not be used to monitor healing, because the imaging findings may lag behind clinical resolution of symptoms.

MRI has more specificity than scintigraphy in distinguishing bone involvement from soft-tissue injuries and may be useful in grading the stage and therefore predicting the time to recovery for certain stress fractures.[56,63] Specific MRI characteristics of stress fractures seen within 3 weeks of the onset of symptoms include (*a*) new bone formation and fracture lines appearing as very low-signal medullary bands that are contiguous with the cortex; (*b*) surrounding marrow hemorrhage and edema seen as low-signal intensity on T1-weighted images, high-signal intensity on T2-weighted images, and short T1 inversion recovery images; and (*c*) periosteal edema and hemorrhage appearing as high-signal intensity on

T2-weighted and short T1 inversion recovery images.[64,65] Although more expensive than scintigraphy, MRI avoids radiation exposure and requires less time than a three-phase bone scintigraphy.[56] Femoral neck stress fracture will sometimes require MRI to better characterize the nature of the fracture and determine nonoperative versus surgical (prophylaxic pinning) treatment.

Computerized tomography may be useful in differentiating a stress fracture from a stress reaction and in correctly diagnosing those conditions with increased uptake on bone scan that may mimic a stress fracture. These include osteoid osteoma, osteomyelitis, and other malignancies.[25]

Fractures that have a propensity for progressing to complete fracture, delayed union, or nonunion are considered high-risk fractures and should be treated more aggressively.[56] Fractures that have been identified as high risk in the general population include fractures of the femoral neck (tension side), the patella, the anterior cortex of the tibia, the medial malleolus, the talus, the tarsal navicular, the fifth metatarsal, and the great toe sesmoids. Tibial stress fractures are common in both men and women; however, women appear to have more femoral, metatarsal, and pelvic stress fractures than men.[22,29,53] Researchers studying 2,962 women undergoing basic training at the Marine Corps Recruit Depot found the most common sites of stress fracture (in descending order of occurrence) to be the tibia, metatarsals, pelvis, and femur.[51] Each will be considered independently.

Pubic Ramus Stress Fractures

Pelvic stress fractures were first described in military recruits in 1937.[66–70] Although relatively rare, representing only 1% to 10% of all stress fractures, they most commonly affect long distance runners[54,71–74] and fencers.[75] In a study of stress fractures occurring in Navy recruits, the rate of diagnosis of pelvic stress fracture was 1 in 367 female recruits, compared with 1 in 40,000 male recruits. Stress fractures of the pubic ramus occur secondary to the chronic pull of the adductor musculature on the thin pubic rami.[67,71,73] Theses stress fractures most frequently occur at the narrowest area of the pubic bone: the isthmus. The isthmus is the site for the origin of the adductor magnus. It has been hypothesized that, as the medial fibers of this muscle act with the adductors to flex the hip and the lateral fibers act with the hamstrings to extend the hip, a shear stress is created that may ultimately produce a fracture.[69] Specific etiological factors include tight adductors, leg-length discrepancy, a crossover running style, and overstriding.[76]

Several studies have demonstrated a relationship between shorter female height and injury rates among military recruits.[4,10,77,78] Those females who developed stress fractures of the pelvis tended to be shorter than average, marched at the rear of their training divisions, and complained of having to overstride while marching.[66] Because shorter soldiers are traditionally assigned to the rear of marching formations, they need to overstride to keep pace with their taller colleagues. This increases the tensile stresses placed on the pubic rami by the adductors and hamstrings, and predisposes these individuals to stress fractures in this area.[66]

On physical examination, it may be difficult to distinguish a pubic ramus stress fracture from an adductor muscle strain. Both diagnoses may present with pain in the groin, tenderness to palpation at the ramus, and pain with resisted adduction of the thigh.[76] An appropriate clinical history and radiographic imaging will likely elucidate the diagnosis.

Femoral Neck Stress Fractures

Although uncommon, the clinician should maintain a high index of suspicion for this particular injury, because there is a high complication rate if the diagnosis is missed or the patient is improperly treated.[79] These fractures are typically from overuse and may develop as the hip musculature becomes fatigued and loses its protective shock-absorbing capability.[56] Coxa vara and osteopenia may also predispose the femoral neck to injury.[56] These fractures present with anterior groin pain that increases with weight-bearing and at the extremes of range-of-motion testing. The pain is often in the distribution of the obturator nerve.[76]

Compression stress fractures begin at the inferior cortex of the femoral neck, are more common than tension-sided stress fractures, and generally heal without complication. Although not necessitating acute surgical treatment, the patient must be closely followed and should remain non–weight-bearing until normal gait without pain. This fracture may take 3 months to heal to become nonsymptomatic.[76] Radiographs may show healing in 6 to 9 months.

Tension stress fractures, on the other hand, start in the superior cortex of the femoral neck and have a propensity to propagate perpendicular across the femoral neck. Because this type of fracture may become displaced, it necessitates aggressive treatment. Complications that may occur after the fracture displaces include delayed union, nonunion, varus deformity, and osteonecrosis.[56] It is therefore recommended that a patient with a tension-sided femoral neck stress fracture be surgically treated with internal fixation.[56,76,80] After surgery, the patient should remain non–weight-bearing for 6 weeks, followed by an additional 6 weeks of partial weight-bearing.[56]

Tibial Stress Fractures

Tibial stress fractures are very common and may occur at any site along the shaft of the bone, but are most frequently encountered in the compression side (posteromedial cortex).[56] Compression-sided stress fractures usually occur in distance runners and respond favorably to cessation of the inciting activity. Athletes who engage in repetitive jumping and leaping activities may develop a stress fracture on the tension side (anterior cortex) of the tibia. This less common—often initially more subtle, yet, ultimately more problematic—stress fracture is predisposed to nonunion, delayed union, and even complete fracture secondary to its relative hypovascularity and constant tension from the posterior muscle forces. The radiographic appearance of this fracture has been referred to as the "dreaded black line" because of its prolonged healing time.[56]

Metatarsal Stress Fractures

Metatarsal stress fractures occur most commonly in the distal half of the second and third metatarsal shafts.[81] In a study of 295 military recruits, it was noted that metatarsal stress fractures characteristically appeared on the medial aspect.[36] The diagnosis is generally straightforward. A patient may experience localized pain and mild edema over the involved metatarsal. Some patients have difficulty localizing their pain and may experience tenderness over adjacent metatarsals. A good way to identify the metatarsal that has the stress fracture is to elicit tenderness by dorsiflexing the digits forcibly. This elicits tenderness of the fractured metatarsal.[82]

Stress fractures of the first metatarsal are uncommon, probably because of its mass and functional ability to transfer load through dorsal motion.[83] Military studies have demonstrated that these fractures occur primarily in the proximal metaphyseal region of the first metatarsal.[82]

Fifth metatarsal stress fractures are also uncommon because of the independent range of motion.[82] They occur at the proximal diaphysis of the bone just distal to the tuberosity and the ligamentous structures.[56] Fifth metatarsal stress fractures have a propensity for delayed union or nonunion and have a high risk of refracture after nonoperative treatment.[56,84]

Shin Pain

The outdated "shin splint syndrome" has been a term used for decades. It describes any condition that produces pain and discomfort in the lower leg, usually because of running, walking, or marching. "Shin splints" is now considered a vague term, and specific injuries should be defined by anatomy, specific site of pain, and cause of injury.[21] Examples of common conditions causing shin pain include exercise-induced compartment syndrome, medial tibial stress syndrome, or periostitis. Shin pain is a common complaint in BCT recruits.[21] In general, intrinsic risk factors for shin pain include previous injury, poor conditioning, long running distances, competitive running, and lack of running experience.[85]

Chronic Exertional Compartment Syndrome

Chronic exertional compartment syndrome (CECS) is one cause of shin pain that may affect women as they train and work in military environments. It differs from acute compartment syndrome, which usually occurs following trauma and is a surgical emergency. CECS occurs when physical activity, especially running, leads to increased pressure in the compartments of the lower leg. This is detrimental to military women, because if they are unable to pass the mandatory physical fitness test, they can be separated from the military. Unfortunately, in patients with chronic compartment syndromes, the average delay in treatment from the onset of symptoms is 22 months.[22]

Incidence/Anatomy. Historically, there has been a low incidence of CECS reported in military medicine and other literature, but it appears to be more prevalent as healthcare professionals are becoming more aware of the condition. Turnipseed[86] reported that 796 of 843 patients with symptomatic lower extremity claudication and without obvious musculotendinous injury or detectable vascular disease were diagnosed with CECS.

With CECS, anterior lower leg pain is brought on with exercise and relieved with rest, although symptoms may persist for several hours. The exact underlying pathophysiology is still not well defined, and there are conflicting reports in the literature about the impact of ischemia in the generation of this symptom complex.[22] One theory is that the exercise-limiting pain occurs because of eccentric muscle contractions, which release proteins within a closed fascial space. These proteins then lead to increased osmotic pressure in the interstitial fluid of the compartment, which likely impairs venous blood egress[87] and reduces capillary blood perfusion. When blood flow to the muscles is decreased, it can lead to tissue ischemia and compromised neuromuscular function. Pain from ischemia and compartment pressure is the primary limiting factor in functional performance.[87] In severe cases, tissue ischemia can lead to tissue necrosis, converting a transient condition into an acute compartment syndrome that requires an emergency fasciotomy.

The anterior compartment is the most commonly affected compartment, followed by the lateral compartment. In a study of young military members with CECS, all 42 participants who were recommended for surgical release had symptomatic anterior compartments.[88] The deep fibular nerve lies in the anterior compartment, and the superficial fibular nerve is found in the lateral compartment. If the pain-producing activity continues, these nerves can become compressed because of the increased pressure in the compartment(s).

Diagnosis. Differential diagnoses include periostitis, fibular or tibial stress reaction or fracture, medial tibial stress syndrome, tenosynovitis, tarsal tunnel syndrome, muscle disorders, popliteal artery claudication, or peripheral nerve entrapment.[87] Other less common possibilities include diabetic peripheral neuropathy, lumbar radiculopathy, spinal stenosis, peripheral vascular disease, infections, and tumors.[89,90]

CECS may be diagnosed by reproducing symptoms with walking or running, compartment pressure testing before and after walking or running, diagnostic imaging, or by utilizing a systematic approach to identify individuals with the condition, as opposed to those with other conditions (eg, tibial or fibular stress reaction or periostitis). The intracompartment pressures can be measured using a wick or slit catheter technique, which are the most popular and reliable methods of assessment.[22] With CECS, elevated resting tissue pressure both before and immediately after exercise has been reported, along with delayed return-to-normal pressure following exercise.[91,92] One of the common criteria used for chronic compartment syndrome was developed by Pedowitz et al[91] and includes any one of the following: (*a*) a preexercise pressure greater than 15 mm Hg; (*b*) a 1-minute postexercise pressure of greater than 30 mm Hg; and (*c*) a 5-minute postexercise pressure in excess of 20 mm Hg. Vath and colleagues[93] presented a case series of young athletic military academy cadets with anterior and lateral leg pain, and suggested it is possible to differentiate CECS from other ailments of the anterior and lateral leg using a prospective assessment of the history, clinical findings before and after treadmill running, and findings from radiographs and MRIs.

Symptoms of CECS include exercise-limiting, moderate-to-severe leg pain that is described as constant with activity and is tight, diffuse, sharp, aching, burning, deep, throbbing, cramping, dull, radiating, and/or squeezing. It can be unilateral or bilateral. Other symptoms may include painful stretching; weakness with running causing the foot to slap; Tinel's to superficial fibular nerve; tenderness with palpation to the anterior and/or lateral compartment; decreased sensation to the leg, dorsum of foot, or first web space; and/or fascial defect(s).[93] The duration of running before onset of symptoms typically varies between individuals. Symptoms usually resolve within minutes after stopping the activity, but may persist longer.

Medial Tibial Stress Syndrome

Incidence/Anatomy. Medial tibial stress syndrome is one of the newer terms for one specific cause of shin pain. As with other previously described overuse injuries, medial tibial stress syndrome can occur if there is a training error, such as a significant change in running or marching distance, surface, intensity, and/or frequency. In a study conducting anthropometric measurements of 17 subjects with and 19 subjects without shin pain, it was found that excessive pronation, excessive velocity of pronation, and poor heel cord flexibility are correlated with shin splints. These factors may lead to the development of medial tibial stress syndrome because of the increased stress and strain to the medial soleus during walking, marching, or running.[94] The medial soleus inserts into the medial one third of the calcaneus and acts as a plantar flexor and an inverter of the ankle. It must work eccentrically to control the calcaneus during the pronation that occurs following heel strike with normal running.[22] Unconditioned runners are especially at risk. In a military study at the US Naval Academy (Annapolis, MD), twice as many of the midshipmen who developed shin splints did not participate in physical training before they began the Academy's rigorous physical training program in comparison with their counterparts who did not develop shin splints.[95] Women are especially vulnerable and may be at higher risk than males.[96]

Diagnosis. Medial tibial stress syndrome usually presents with moderate increase of radionucleotide activity along the posteromedial border of the tibia with bone scintigraphy, in contrast to the typical focal and intense reactions seen with tibial stress fractures.[22] Routine plain radiographs of the tibia are often reported as normal, but there can be hypertrophy of the posterior tibial cortex from the bone remodeling as it adapts to increased repetitive stress.[22] Tenderness is common along the posteromedial border of the tibia, usually from approximately 4 cm proximal to the medial malleolus and extending proximally for a variable distance up to 12 cm. Michael and Holder[97] investigated cadavers to correlate clinical findings with electromyographic and bone scanning results to conclude that tenderness directly over the posteromedial border of the tibia corresponded to the medial origin of the soleus muscle. The pain from medial tibial stress syndrome usually begins as a dull ache or soreness and can advance to a sharp, penetrating, and severe pain as the condition worsens. It can eventually progress to discomfort during daily activities. In addition, there may be slight swelling and pain with active-resisted plantar flexion and calf raises.[22] This is especially true if the calf raises are conducted with the knee in slight flexion, putting the gastrocnemius on slack.

Common Overuse Problems of the Knee

Patellofemoral Syndrome

Incidence/Anatomy. Patellofemoral syndrome (PFS) is often categorized as anterior knee pain, but the entity is more complex than this. In general, conditions causing patellofemoral symptoms include patellofemoral pain syndrome, tracking disorders, instability or dislocation, and chondromalacia.[98] Pain is usually in the peripatellar or retropatellar area, and is increased with repeated or prolonged deep knee flexion that occurs with climbing stairs or sitting for prolonged periods. It is more common in females and has been associated with various limb malalignments. These include increased femoral anteversion, external tibial torsion, foot pronation, an increased Q angle (the angle formed by intersection of a line drawn from the anterior superior iliac spine to the center of the patella and from the latter point to the tibial tuberosity),[99] and patella alta.[76] Additionally, PFS is associated with ligamentous laxity. Although these anatomical variations are not limited to women, they are seen more commonly in the female population.[76]

Diagnosis. The diagnosis is generally made by history and physical examination alone. As noted previously, patients complain of pain over the anterior knee or patella (peripatellar or retropatellar). In addition, patients may report a history of patellar dislocation that required reduction or was even spontaneously reduced, but this should be differentiated from "buckling" or giving way. This dislocation may also be associated with a history of trauma or twisting.[100] This differentiation is important because these symptoms generally point to patellar instability (sometimes known as patellar subluxation), as opposed to PFS. The etiology of pain without instability is more difficult to differentiate.

An understanding of the nature of patellar pain should also be sought. Specifically, constant pain that is not activity related may be referred, as well as radicular pain, sympathetic-mediated pain, or secondary-to-postoperative neuroma.[100] Sharp, intermittent pain or associated signs like catching or locking may be secondary to intraarticular pathology, such as loose bodies, meniscal tears, or a plica syndrome (inflammation or thickening of synovial folds). Activity-related pain may be caused by articular tissue overload secondary to chronic patellar malalignment or soft-tissue overload without patellar malalignment. Examples of the latter include patellar or quadriceps tendinitis, fat pad syndrome, ITB syndrome, or early lateral patellar compression syndrome.[100] Inflammatory arthritides, myalgias, or generalized deconditioning may also manifest as anterior knee pain.

Iliotibial Band Friction Syndrome

ITBFS is a fairly common cause of lateral knee pain secondary to overuse in military personnel.[101] It was first reported by Colson and Armour[102] in 1961. It is also seen in long distance runners, weightlifters, downhill skiers, and cyclists, but there is no known predilection in females.[103–106] The exact etiology, however, remains somewhat controversial, and the diagnosis is based almost solely on history and physical examination. Management is almost always conservative.

Incidence/Anatomy. Depending on the population studied, ITBFS has been reported in up to 50% of various groups.[101] Renne[105] reported 16 cases in 1,000 Officer Candidate School trainees in 1975. In addition, Linenger et al[107] considered this to be the most common injury in Marine basic training. Anatomically, the ITB is formed by the coalescence of the tensor fascia lata, gluteus maximus, and gluteus minimus fascia proximally; continuing distally as an attachment along the linea aspera of the femur; and inserting distally on the tibia and patella. It is this distal attachment that is of particular interest. Specifically, it has two tracts: (1) the iliopatellar tract that influences deceleration and prevents medial patellar subluxation; and (2) the iliotibial tract that attaches to Gerdy's tubercle and the biceps tendon, and functions as an anterolateral ligament, providing lateral knee stabilization.[108,109] It is here, between the ITB and the lateral femoral condyle where the bursa sits, that repetitive friction can start an inflammatory process. Several authors have even identified a fluid collection or inflammation deep to the ITB, with ITB thickening on MRI in patients with clinical symptoms consistent with ITBFS.[110–112] However, other authors have considered the area more consistent with the lateral synovial recess or lateral extension of the joint capsule.[113] In anatomical dissections, Nemeth[113] found this area to have chronic inflammation, hyperplasia, fibrosis, and mucoid degeneration, thus lending support to the extension of the joint capsule concept. Nonetheless, it still appears possible that a bursitis could coexist.

Functionally, the ITB is anterior to the lateral femoral condyle when the hip is flexed. With subsequent knee flexion of 30 degrees or more, the gluteus maximus contracts and the ITB moves posteriorly. Friction, however, occurs in the early stance phase[101] and is increased in some individuals with repetitious movement. Factors that have been suggested as contributing to ITBFS include excessive mileage, hill training, abnormal foot mechanics, a tight ITB, genu varum, a prominent lateral femoral condyle, and leg-length discrepancy.[103–106,114–117] Training errors have most commonly been implicated, even after a single severe session.[118] With sudden increases in training distance, improperly fitting or designed training shoes or constantly running on sloped roads may also contribute. Nonetheless, some of the proposed etiologies or contributing factors have never been proven, and most likely the true etiology is multifactorial.

Diagnosis. As previously described, ITBFS is almost exclusively diagnosed by history and physical examination. Pain is usually reported over the lateral femoral condyle, and, consequently, should be differentiated from other common causes of lateral knee pain, such as ligamentous strain, lateral meniscal tear, patellofemoral instability or maltracking, osteochondral injury, tendonitis, or simply arthritis.[101,119] Symptoms frequently increase with running or stairs secondary to excessive knee flexion with these activities.

On physical examination, knee instability, point tenderness, and effusion should be evaluated. ITB tightness can be assessed with Ober's test,[120] whereas single leg standing may elicit a stinging pain over the lateral femoral condyle (Renee creak test).[105] Direct pressure on the lateral femoral condyle with the patient in the supine position and the examiner extending the knee may also elicit discomfort (Noble test).[117]

As noted previously, MRI is limited, although a negative MRI may help to differentiate the clinical symptoms from other etiologies.[110–112] Occasionally, a fluid collection deep to the ITB or ITB thickening has been noted. In addition, technetium bone scanning has shown increased radiotracer uptake over the posterolateral lateral femoral condyle; nonetheless, this should not routinely be obtained as a diagnostic tool for this condition.[121]

Patellar Tendinitis/Tendinosis

Incidence. The patellar tendon is one of the most common sites of tendinitis occurring in BCT.[21] Activities such as marching or running downhill, jumping off obstacles, performing jumping jacks, or playing sports can lead to patellar tendinopathy in the military. In the general population, volleyball and soccer are the most common sports involved in patellar tendinopathy,[122] and the greatest incidence occurs between the late teen years into the 30s.[22] Patellar tendinitis and tendinosis are commonly referred to as "jumper's knee" because repetitive jumping can lead to tendinopathy. However, other causes of patellar tendinopathy include those that produce eccentric quadriceps muscle contractions, such as marching downhill, descending stairs, or lowering a weight.[21]

Diagnosis. Patients will complain of anterior knee pain typically with eccentric loading of the quadriceps muscle. Acutely, there will be pain with resisted knee extension and tenderness to palpation over the distal patellar pole and the proximal portion of the patellar tendon.[22] The tendon itself may be edematous and feel "boggy" on examination.[22] The patient will most likely be symptomatic during bilateral or single-leg hop testing.

Common Overuse Problems of the Ankle and Foot

Achilles Tendinitis or Tendinosis

Incidence/Risk Factors. Achilles tendinitis and tendinosis are common lower extremity overuse conditions occurring in military women. Achilles injuries can be caused by intrinsic or extrinsic factors, or a combination of both. Intrinsic factors may include aging and degeneration of the tendon, decreased vascularity, heel-forefoot malalignment, poor gastrocnemius-soleus flexibility, excessive dorsiflexion range of motion, and foot hyperpronation. If the Achilles is tight, hyperpronation will be exaggerated and there will be further stress on the tendon.

Extrinsic factors may include sudden increases in training intensity or duration, change of running surfaces from soft to hard, inappropriate or old footwear, and repetitive forces such as marching/running up and down hilly terrain or performing jumping jacks. These can all lead to overload of the Achilles tendon, causing acute or chronic tendinitis and possibly tendinosis because the repetitive forces do not allow full recovery between applied stresses. There is an area of avascularity in the Achilles tendon that is located 2 to 6 cm above its insertion into the calcaneus, and this area is vulnerable to inflammatory changes and degenerative changes within the tendon. This site is also vulnerable to Achilles tendon ruptures, especially if repetitive microtrauma to an Achilles tendinopathy occurs over time, and the tendon is suddenly and aggressively stretched while simultaneously contracting eccentrically.[87]

Women who are not physically fit, who do not have much running experience in the past, who have weak ankle plantar flexor muscles, or who have increased dorsiflexion range of motion may be the most vulnerable for Achilles injuries. Their Achilles tendons may not fully adapt to the increased functional demands of running, marching, and/or jumping, and are susceptible to injury. Changes in force demand on tendons may not allow tenocytes to either repair the initial damage or fully adapt to the new loading state.[123] Continued training under these conditions can lead to further tissue damage and progressively greater functional deficits as the collagen becomes denatured and cross-links break.[87]

Diagnosis. Clinical signs and symptoms of Achilles tendinitis and/or tendinosis include antalgic gait; tenderness, swelling, and/or palpable nodules in the area 2 to 6 cm proximal to the calcaneal insertion; pain during heel raises; and painful stretching to the Achilles tendon. If there is exudation around the tendon—commonly found in the acute stage—crepitus may be present, and is accentuated by active dorsiflexion and plantar flexion of the foot.[22] In addition, there may be decreased dorsiflexion range of motion and/or hyperpronation. Pain is the most common symptom initially and usually lessens with rest, but can be exacerbated by climbing stairs. In the subacute stage, pain occurs at the beginning of a run and becomes worse with sprinting, sometimes forcing the individual to stop. During the advanced stages, when there is tendinosis or a partial rupture, running is not possible and pain occurs at rest.[22]

Posterior Tibialis Tendinitis/Tendinosis

Incidence/Anatomy. The posterior tibialis tendon is the most anterior structure at the level of the medial malleolus and is located immediately adjacent to the posterior surface of the malleolus.[22] "It is the main dynamic stabilizer of the hindfoot against valgus (eversion) forces owing to its multiple insertions."[22] Specifically, the posterior tibialis muscle works eccentrically to provide control of subtalar joint motion during walking, marching, and running. If there is abnormal subtalar joint pronation, there will be increased posterior tibialis muscle activity, which can increase the risk of overuse injury (eg, tendinitis/tendinosis). Running on a crowned road is an example of an extrinsic cause of excessive pronation.

Posterior tibial tendon disruptions can occur if there is a long history of posterior tibialis tendinosis/tenosynovitis, especially in women over age 40. The classic clinical signs of a posterior tibial tendon are hindfoot valgus, adduction of the forefoot, a positive single heel rise test (calcaneus does not invert), a severe navicular drop sign, a "too-many-toes" sign (as viewed from behind), and weakness of the tendon.[22]

Diagnosis. Those suffering from posterior tibialis tendinitis/tendinosis may feel pain at the medial posterior ankle, especially if they must run or perform any activity requiring a strong pushoff action. This condition can be caused by pes planus/hyperpronation; weakness or poor endurance of the posterior tibialis muscle; tight gastrocnemius-soleus complex; and/or stiffness in the tibiotalar, talonavicular, and calcaneocuboid joints. Posterior tibialis tendinitis or tendinosis is usually characterized by tenderness and/or swelling to the distal tendon proximal to the navicular tuberosity. There may be tenderness on the navicular tuberosity and/or posterior to the medial malleolus, along with crepitus over the tendon on active movement. Pain also occurs with resisted ankle plantarflexion and inversion. The primary differential diagnosis to consider is tarsal tunnel syndrome.

Plantar Fasciitis

Incidence. Plantar fasciitis is another common lower extremity overuse injury occurring in military women. It may be associated with wearing boots in BCT, but no evidence exists to demonstrate this as a true risk factor.[21] There has been research and much discussion about the relationship of calcaneal heel spurs and subcalcaneal pain, but this relationship has not been definitely established.[22]

The foot's windlass mechanism as described by Hicks[124] occurs when the arch is raised and supported with dorsiflexion of the toes, providing more stability to the foot. Those with pes planus (hyperpronation) or pes cavus appear to be more vulnerable to plantar fasciitis, perhaps because both foot types place extra stress on the plantar fascia.[87] A pes planus foot will place increased pull on the origin of the plantar fascia at the calcaneus because the windlass mechanism will be under increased strain while maintaining a stable arch during the pushoff phase of gait.[22] A pes cavus foot will cause excessive strain at the heel insertion because the rearfoot is unable to adequately evert to absorb the shock and adapt itself to the ground.[22]

Diagnosis. Plantar fasciitis begins as insidious onset of pain that originates in the plantar heel and is usually worse in the morning right after getting out of bed. There is tenderness with palpation over the medial calcaneal tubercle, occasional swelling over the plantar medial aspect of the heel, and possible antalgic gait.[87] Occasionally, the medial calcaneal nerve is inflamed and entrapped, and will cause paresthesias or pain in the area.

Common Traumatic Injuries in Military Women

Traumatic injuries are typically "due to sudden energy exchanges resulting in abrupt overload with tissue trauma."[20] In a study of Army wheel vehicle mechanics, traumatic injuries accounted for 32% of female injuries, compared with 49% of male injuries.[20] This generally agrees with a previous study looking at traumatic injuries at Ft Lewis, WA, wherein military men were 1.2 times more likely to be injured than women.[125] However, women still experience traumatic injuries. Examples of traumatic injuries in military women include anterior cruciate ligament (ACL) tears, ankle sprains, acetabular labral tears, lumbar strains, and sacroiliac joint (SIJ) dysfunctions.

ACL Ruptures

Incidence/Anatomy. ACL is the primary restraint to anterior shear and internal rotation of the tibia on the femur. It is the most frequently ruptured knee ligament, especially among female athletes, where the rates of ACL ruptures for female athletes range from 2.4 to 9.7 times higher than in male athletes.[126] Seventy percent of ACL injuries are noncontact and typically occur during deceleration, pivoting, landing, or responding to a perturbation.[127]

Risk Factors. A retreat was conducted in Lexington, KY, in 2003 where more than 50 physicians, physical therapists, athletic trainers, and scientists from around the world—specializing in the areas of biomechanics, motor control, and neuromuscular function—attended to formulate a consensus statement focusing on the gender bias in ACL injuries.[128] The consensus statement includes information that is both known and still unknown regarding structural, neuromuscular, biomechanical, and hormonal factors, along with interventions.[128]

According to the consensus statement, known *structural factors* include the following: (*a*) females have wider pelvis-to-femoral length ratios that can contribute to genu valgum; (*b*) females have larger Q angles; (*c*) "the size and shape of the notch may contribute to stenosis and ACL impingement and injury"; (*d*) "the combination of knee abduction (valgus) and external rotation positions contribute to ACL impingement in vitro, especially in the presence of a narrow notch—these motions have been shown to be greater in females, compared to males during athletic activity"; and (*e*) "females have greater laxity and greater active hip rotation range of motion compared to males."[128]

Known *neuromuscular factors* include the following: (*a*) females activate muscles earlier than males in anticipation of landing; (*b*) increases in knee joint loading during movements linked to noncontact ACL injury can be caused by variability in neuromuscular control parameters at impact; (*c*) females rely more on quadriceps to stabilize the knee; (*d*) females have reduced muscle stiffness when trying to control knee motion; (*e*) females take longer to produce quadriceps and hamstring muscle tension with reflex activation after fatiguing exercise with isokinetic testing; and (*f*) females demonstrate lower muscular endurance, which can increase risk of injury.[128]

Known *biomechanical factors* include the following: (*a*) mechanisms for noncontact ACL injuries include deceleration with the knee in an extended position, landing from a jump, and sidestep cut maneuvers; (*b*) females have more knee valgus during sidestep cutting and more knee extension at initial contact; (*c*) "compared to running, frontal- and transverse-plane movements are greater during anticipated sidecut maneuvers to 30 degrees and 60 degrees and crossover cutting—in addition, these moments were increased when the tasks were performed under unanticipated conditions"; (*d*) females exhibit less knee flexion and increased hip and knee internal rotation during single-leg landing and forward-hopping; and (*e*) females have greater knee extension and valgus moments during landing phases of jump stop tasks, which are associated with increased ACL strains.[128]

Known *hormonal factors* include the following: (*a*) self-report of menstrual phase is not accurate; (*b*) in rats, there is no change in ultimate failure load of the ACL throughout the estrous cycle; (*c*) physical performance may be influenced by hormone-dependent changes throughout the menstrual cycle; (*d*) in ovariectomized rabbits, supraphysiological levels of estradiol decrease the load-to-failure; (*e*) individual variations in sex hormone concentration is associated with changes in low-load viscoelastic properties of the tibiofemoral joint; and (*f*) during pre- and postexercise, there are differences in knee joint laxity between genders, but there is a similar increase in knee joint laxity postexercise.[128]

Lateral Ankle Sprains

Incidence. The most common sprain in BCT and AIT is the lateral ankle sprain.[21] Another study demonstrated that, among female Marine recruits, the most commonly reported injuries were PFS (10.0% of subjects), ankle sprains (9.1%), and ITB syndrome (5.8%).[129] Ankle sprains are also prevalent in young athletic populations, such as in military academies. Jackson et al[130] reported that 33% of the cadets at the US Military Academy sustained an ankle sprain during their 4 years of schooling. At the same institution, Gerber et al[131] evaluated all cadets presenting with ankle injuries during a 2-month period and noted that, of the 104 ankle injuries (which accounted for 23% of all injuries seen), there were 96 sprains. Seventy-nine percent of the sprains were lateral ankle sprains.[131] Similarly, others have reported that 85% of all athletic-related ankle injuries are lateral ankle sprains.[87]

Common causes of lateral ankle sprains in a military environment include running or marching in formation, parachuting, participating in physical training and sports, and training on obstacle courses. Marching or running in formation is a high-risk factor, especially if performed in the middle of the formation when obstacles are not always seen or anticipated, on uneven terrain, or in poorly lit circumstances.

Parachuting is another high-risk activity for ankle sprains in the military. Twenty-six percent to 41% of airborne operations injuries involve the ankle.[20,132–135]

Sports are a common form of military physical training and team morale building. US Military Academy cadets participate in extensive physical fitness and competitive sports on a regular basis. Sports such as soccer, basketball, and football that require jumping and cutting are frequently the cause of lateral ankle sprains. Throughout the year, cadets engage in basketball, rugby, running, and many other athletic activities. In the fall season, more than 600 cadets play either intramural or intercollegiate tackle football, and a similar number participate in soccer, offering numerous opportunities for ankle sprains.[131] Studies in the civilian athletic population demonstrate that women have a higher risk of sustaining ankle sprains than men in some sports. In a prospective study evaluating the relative risk of ankle injuries in scholastic and collegiate basketball players, 11,780 athletes (4,940 females) participated. Overall, females had a 25% greater risk of sustaining a grade I ankle sprain compared with the men.[136]

Mechanism of Injury/Diagnosis. The lateral ankle is sprained more frequently than the medial ankle because of the anatomy and mechanism of injury. Plantar flexion and subtalar inversion will cause a lateral ankle sprain if too much of the body's weight is rolled over the ankle. The number of ligaments involved and severity of the injury will depend on the amount of stress applied and the position of the talus. Sprains are graded based on the magnitude of the tear and the subsequent joint instability:

- *grade I*: a small disruption to the ligament without any joint instability or opening of the joint during a stress maneuver—it is characterized by microfailure of collagen fibers within the ligament[21];
- *grade II*: a partial tear with some instability and partial opening of the joint during a stress maneuver; and
- *grade III*: a complete ligament tear with full joint opening during a stress test.

The anterior talofibular ligament is the most commonly injured ligament, but the calcaneofibular ligament and then the posterior talofibular ligaments can be damaged if the ankle remains inverted while it moves from plantar flexion toward more and more dorsiflexion.[87] An important differential diagnosis to consider would be a syndesmotic injury (high ankle sprain) or a Lisfranc injury, which will require considerably different management.

Acetabular Labral Tears

Incidence/Anatomy. Acetabular labral tears are becoming more recognized as a source of anterior hip or groin pain, especially with recent advances in magnetic resonance arthrograms and hip arthroscopy techniques. Females suffer from acetabular labral tears more often than males.[137–140] The acetabular labrum is a fibrocartilage and dense connective tissue ring attached to the bony rim of the acetabulum. It is thought to aid in the stability of the hip joint by deepening the hip joint and reducing contact stress by increasing the contact area and distributing the load.[141] A tear of the acetabular labrum is believed to result from excessive forces at the hip joint,[139] such as slipping, falling, excessive hip external rotation and/or hyperabduction, running, or twisting. Specific sports/recreational activities that may place excessive forces at the hip include soccer, hockey, golf, and ballet.[141] Researchers have demonstrated that 22% of athletes with groin pain have hip labral pathology.[142] Although these particular activities and movements are thought to cause labral tears, up to 74.1% with mechanical hip symptoms are not associated with a specific event or cause.[13,140,143]

Mechanisms of Injury/Diagnosis. Acetabular labral tears occur in a military population and can be the cause of groin and/or lateral hip pain, catching, clicking, popping, snapping, and/or clunking. In a military tertiary medical center, 21 participants aged 18 to 47 (mean 31.4 ± 8.1 years) were enrolled in a prospective study evaluating clinical assessment, magnetic resonance arthrograms, arthroscopic treatment, and the clinical outcomes of symptomatic acetabular labral tears.[143] Mechanisms of injury included idiopathic injury, fall from heights, childbirth, and pivoting or twisting. A traumatic mechanism of injury was identified in 33% of the patients. A twisting injury was identified in 14% of the patients, with the remaining 53% reporting an insidious onset. Most patients reported pain in more than one location around their hip. Mechanical symptoms occurred in 86% of the patients. The average loss of motion of the affected hip was as follows: 12 degrees of abduction, 9 degrees of external rotation, 8 degrees of flexion, 4 degrees of adduction, and 3 degrees of both extension and internal rotation loss. Common physical examination findings include pain with resisted straight-leg raise (especially when the hip is externally rotated) (Figure 1), pain and/or popping with internal

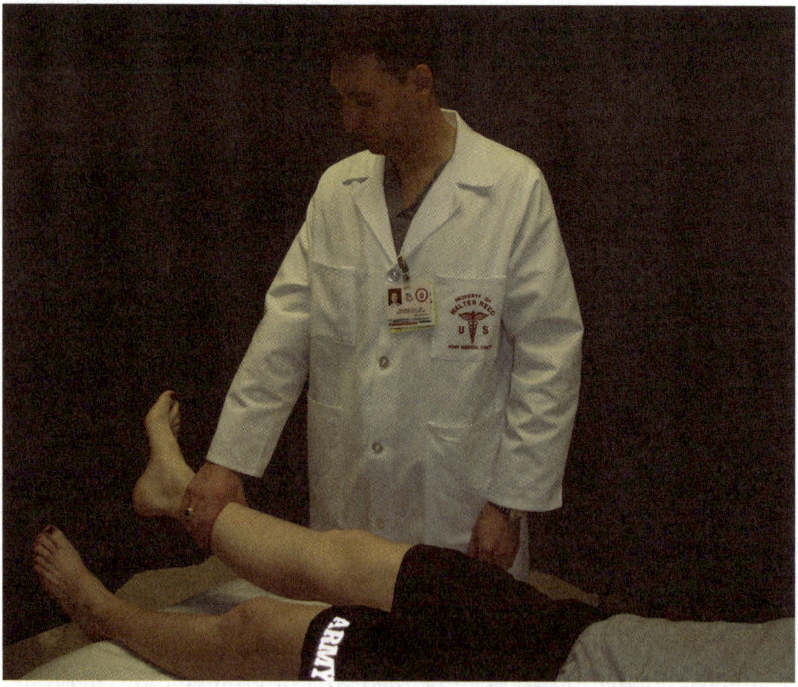

Figure 1. *Resisted straight leg raise in external rotation.*

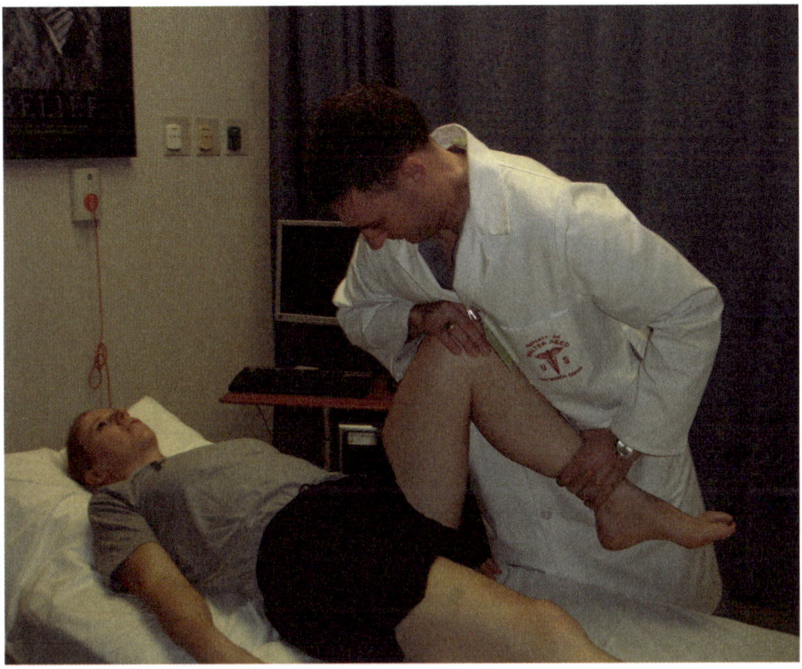

FIGURE 2. *Internal rotation load grind.*

rotation load and grind (Figure 2), and positive tests as described by Fitzgerald[137] and McCarthy et al.[139] A positive McCarthy sign (also known as the Thomas sign) is when both hips are fully flexed, then pain is reproduced by extending the affected hip, first in external rotation and then in internal rotation (Figures 3–5). The Fitzgerald test is a two-part test described by Fitzgerald.[137] To check for anterior labral tears: bring the hip into full flexion, external rotation, and full abduction; then extend with internal rotation and adduction (Figures 6 and 7). To check for posterior labral tears: move the hip into extension with abduction and external rotation from the fully flexed, adducted, and internal position (to check for posterior labral tears) (Figures 8 and 9). These are considered positive if sharp pain is reproduced with or without an associated click.

Back Injuries

Military women suffer from back problems throughout their military careers. Examples of common conditions include lumbosacral strain, intervertebral disc syndrome, and SIJ dysfunction.

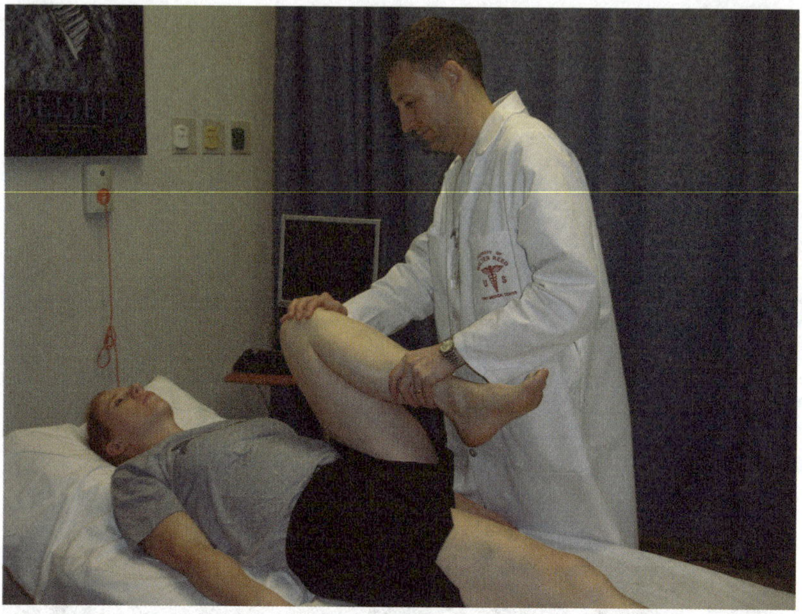

Figure 3. *McCarthy sign in starting position.*

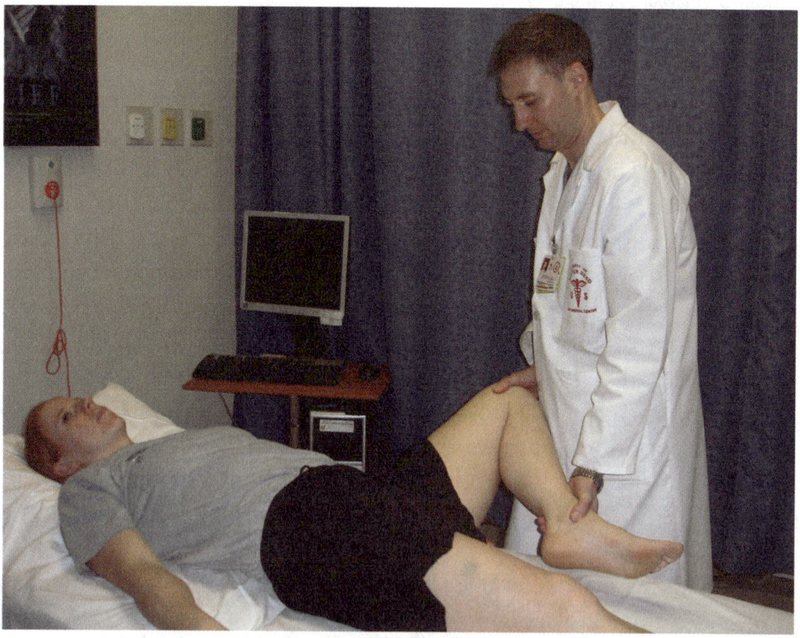

Figure 4. *McCarthy sign going into external rotation.*

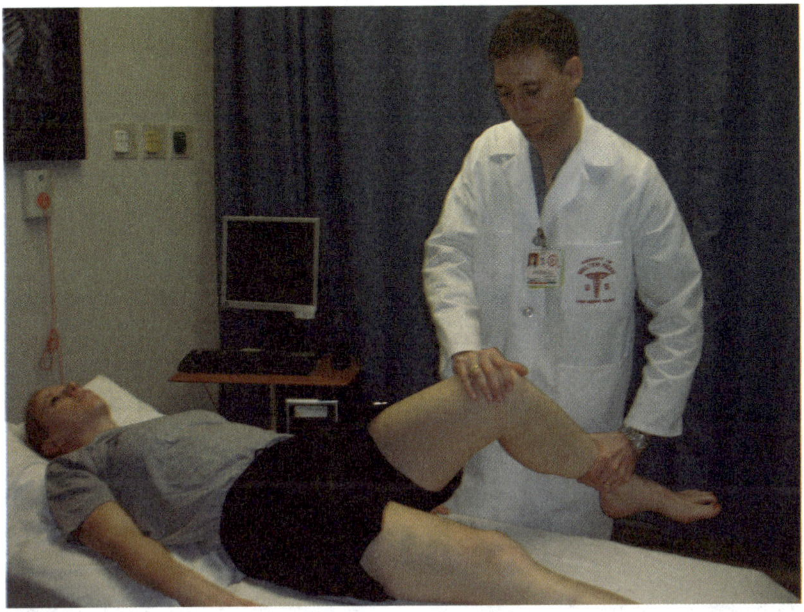

FIGURE 5. *McCarthy sign going into internal rotation.*

Lumbosacral Strain and Intervertebral Disc Syndrome

Acute overuse strains are the most common causes of low back pain during BCT or AIT.[21] However, chronic back conditions are also problematic and can be reason for disability compensation after military service. Orthopaedic conditions of the spine and knee are associated with the highest percentages of total disability cases.[1] A study analyzed 41,750 disability cases using the US Army Physical Disability Agency (USAPDA) database (data included from 1990 to 1994) to determine the following: (*a*) the prevalence of work-related musculoskeletal disability, (*b*) if certain occupations were associated with a greater risk of occupational back disability (OBD), and (*c*) if certain occupations were associated with greater risk of OBD in women.[144] Results showed that, among US Army personnel, lumbosacral strain and intervertebral disc syndrome were the most prevalent diagnoses for back disability cases (67% of all back-related diagnoses and more than 13% of all physical disability diagnoses).[144] Women experienced higher rates of all physical and musculoskeletal disabilities when compared with men.[144] The top five MOSs associated with the highest female OBD (and in which females experience substantially higher relative risk of OBD than males in the

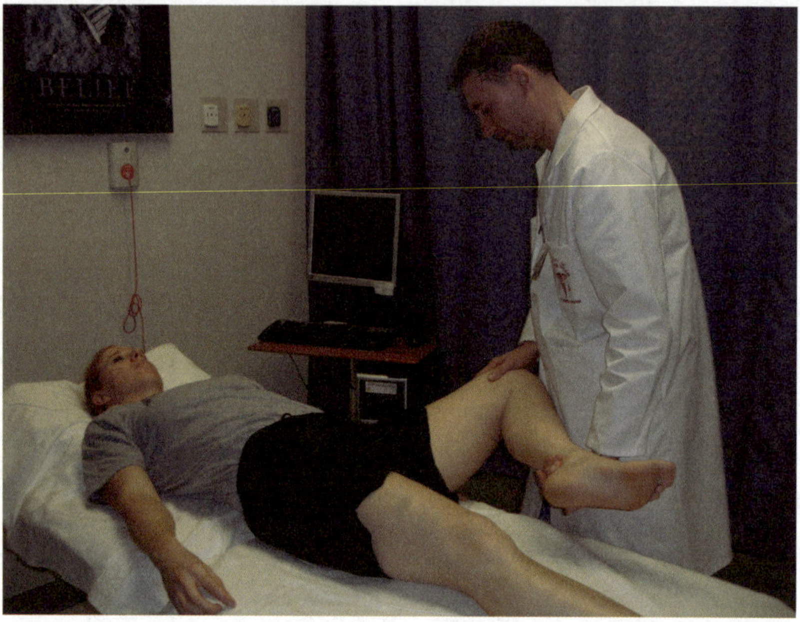

Figure 6. *Fitzgerald test for anterior labral tears—beginning.*

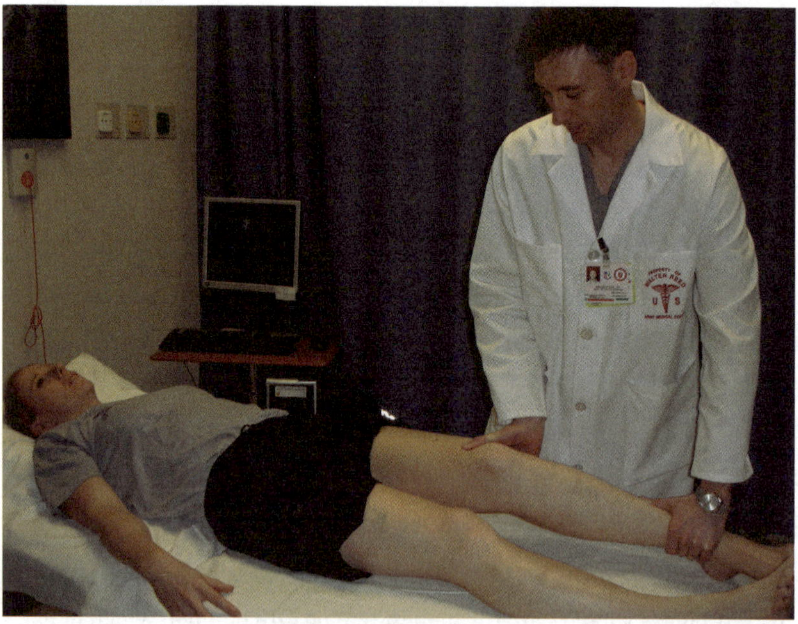

Figure 7. *Fitzgerald test for anterior labral tears—near end.*

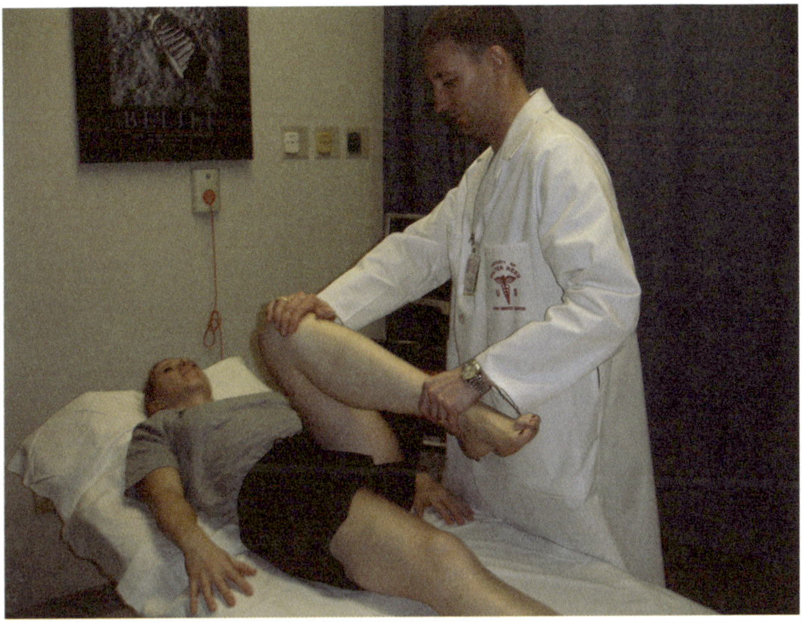

Figure 8. *Fitzgerald test for posterior labral tears—beginning.*

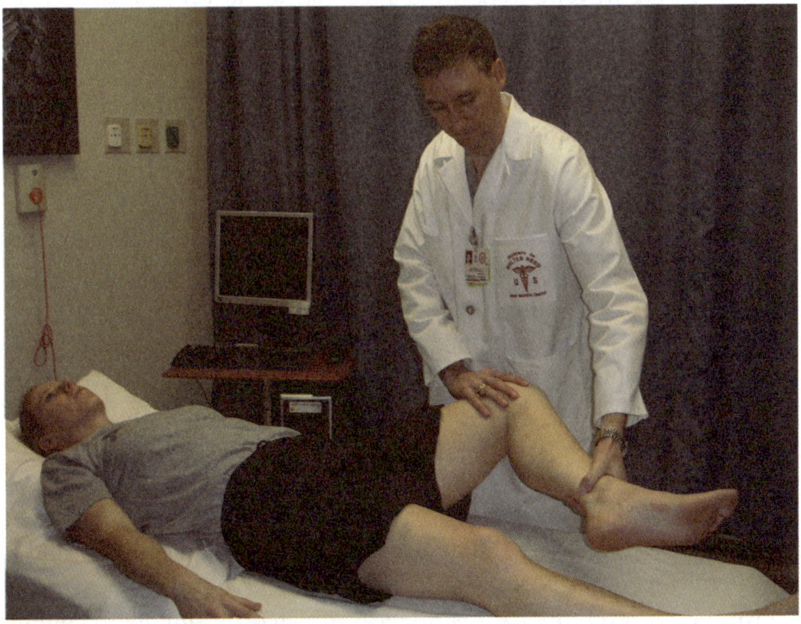

Figure 9. *Fitzgerald test for posterior labral tears—near end.*

same jobs) were the following: (1) multichannel transmission systems operator, (2) single-channel radio operator, (3) wheeled-vehicle mechanic, (4) signal intelligence analyst, and (5) voice interceptor.[144] These jobs tend to be moderately or very physically demanding and require awkward positions: lifting and carrying heavy equipment, bending and twisting, pushing and pulling, and repetitive strain. It is possible that some low back disability cases were a consequence of early training injuries.[144] Specifically, in Army wheeled-vehicle mechanics, 10% of the injuries were in the lower back for women.[20]

An additional study used the same database to specifically determine (*a*) the prevalence of work-related back disability diagnoses; (*b*) specific jobs associated with greater risk of back disability; and (*c*) association among gender, job type, and disability.[145] The specific jobs that were identified in which females sustained higher rates of back disability than males include unit-level, wheeled-vehicle mechanic; single-channel radio operator; multichannel transmission systems operator; interrogator/translator; and practical nurse.[145] These findings are very similar to the previous results.[144]

Sacroiliac Joint Conditions

Incidence. SIJ conditions are challenging and affect 10% to 25% of patients with low back pain.[146–149] Furthermore, 20% to 80% of pregnant women experience low back pain or pelvic pain that often comes from the SIJ.[149–153] Women aged 15 to 40 years appear to be most vulnerable. This is possibly because of hormonal factors and subsequent increased mobility. Military women are susceptible to SIJ dysfunction not only because of the increased prevalence during pregnancy, but also because of the nature of the training and jobs. SIJ dysfunction is associated with a history of minor direct trauma, such as falling on the buttocks or pushing a heavy object.[149,154] Military activities that are potential contributors to SIJ dysfunction include jumping off vehicles or obstacles, landing hard on one leg, stepping into a pothole, climbing, falling on buttocks, lifting and twisting, pushing, marching or running with leg-length discrepancies, or performing activities requiring single-leg stance (eg, on obstacle courses).

Diagnosis. Patients with SIJ dysfunction usually present with a pain referral pattern that includes an area extending 10 cm caudally and 3 cm laterally from the posterior superior iliac spine.[148] It does not extend above the L5 level. SIJ pain is often aggravated by running, climbing stairs, getting out of a car, or any activities that require asymmetrical loading (eg, golfing).[149] The pain can also radiate into the buttocks, groin, and lower limb because there are various sources and levels of innervation ranging from L2 to S2, and the diagnosis may be complicated by discogenic pain or facet joint arthritis.[149] Differential diagnoses include ankylosing spondylitis, Reiter's syndrome, psoriatic spondylitis, infection, tumor, metabolic disorders, degenerative disease, iatrogenic conditions, and referred pain.[149]

Physical examination and clinical tests for SIJ dysfunction include examining the hips, spine, and pelvis. Typically, the physical examination and history are enough to diagnose SIJ dysfunction; but, if physical examination is inconclusive, the next step in diagnosis is often plain radiographs. If that is negative, then a bone scan can identify areas of increased activity in the bone. This is a nonspecific test and can be positive in cases of arthritis, infection, and fracture or tumors of the bone. Point tenderness over the sacral sulcus is one of the most consistent physical findings in those with SIJ dysfunction.[149] Other clinical tests include static and movement symmetry tests, and pain provocation tests. However, many of the symmetry tests have variable reliability.[155,156] The pain

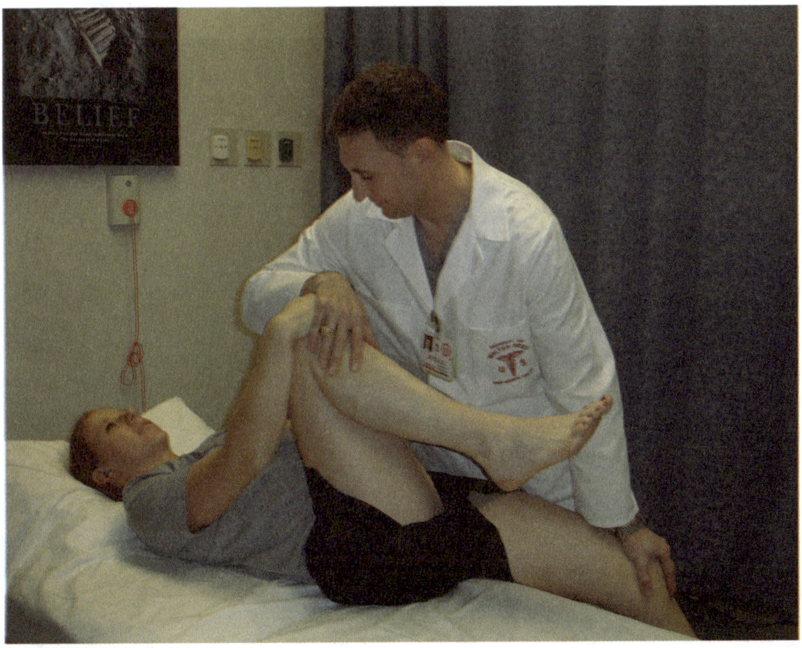

FIGURE 10. *Gaenslen test.*

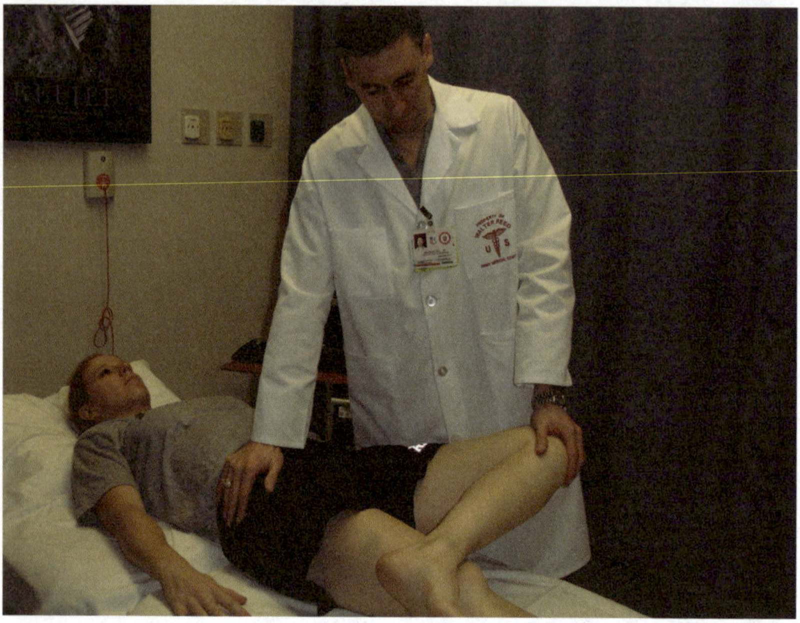

Figure 11. *FABER test (flexion abduction and external rotation).*

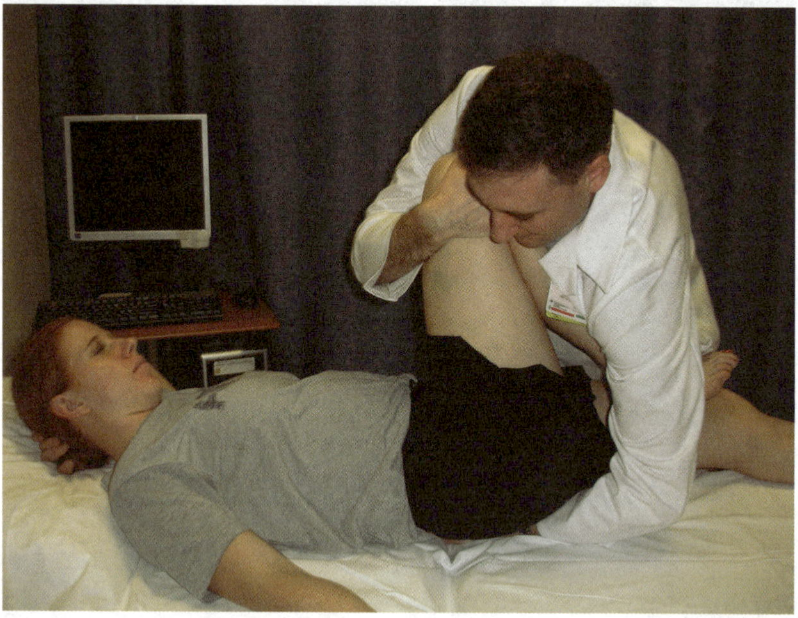

Figure 12. *POSH test (posterior shear).*

provocation tests are the most reliable. The Gaenslen, FABER (flexion abduction and external rotation), and POSH (posterior shear) tests are the most reliable techniques, with an acceptable clinical reliability of greater than 80%.[155,157] The Gaenslen test is performed with the patient supine on the table, and one hip is maximally flexed while the other is maximally extended simultaneously to stress both SIJs (Figure 10). The FABER test starts with the patient supine and hip in flexion, abduction, and external rotation with the lateral ankle on the knee. Overpressure is applied to the medial knee while stabilizing the opposite side of the pelvis (Figure 11). The POSH test is performed with the patient supine and the hip to 90 degrees of flexion and in slight adduction, and axial pressure is applied along the femur (Figure 12). Only the POSH test has acceptable validity with sensitivity and specificity greater than 80%.[158,159] Because there is limited reliability and validity of individual clinical SIJ tests, clusters/combinations of SIJ tests have been studied and shown to be more sensitive and specific.[158,160]

Prevention, Treatment, and Rehabilitation

Overview of General Injury Prevention, Treatment, and Rehabilitation Techniques in the Military

Extrinsic Risk Factors

Prevention and/or rehabilitation of musculoskeletal injuries in military women will depend on the cause: whether it's extrinsic, intrinsic, or a combination. Extrinsic risk factors are external, or environmental, and may include training variables and running surfaces. Extrinsic factors in BCT have been identified and include "high running mileage, training company, older running shoes, and the summer season."[21] The training company is a risk probably because there are variations in the physical training intensities.[19] Higher temperatures may be associated with a higher overall rate of injury in the summer.[1]

Intrinsic Risk Factors

Intrinsic risk factors are personal characteristics and may include variables such as age, height, weight, gender, and physical fitness. Intrinsic "risk factors in BCT include female gender, high foot arches, knee Q angle > 15 degrees, genu valgus, past ankle sprains, low aerobic fitness, low muscular endurance, high and low extremes of flexibility, low levels of physical activity prior to BCT, cigarette smoking prior to BCT, and older age."[21] The strongest evidence (supported with five or more studies) suggests that female gender, low aerobic fitness, high and low extremes of flexibility, low levels of physical activity prior to BCT, and cigarette smoking prior to BCT are risk factors.[1] Risk factors specifically for female medics in AIT include "higher body weight, older age, and a break in training between BCT and AIT (approximately 9 months)."[21] Psychosocial factors such as job/assignment and life satisfaction, especially in the cases of low back pain, should also be considered[145] because they are important predictors for soldiers discharged from the US Army with disability related to occupational low back pain.[1]

General Prevention Strategies

Once the risks are identified, prevention and intervention strategies need to be applied. Unit commanders, noncommissioned officers responsible for unit physical training, unit master fitness trainers, medical officers, medical treatment facility commanders, subject matter experts (physical therapists, dieticians, etc), preventive medicine officers, and service members all need to be involved. Technical Bulletin Medical 592 ("Prevention, Control, and Management of Musculoskeletal Injury Associated with Army Physical Training") is an excellent source of information.[1]

Most of the injury prevention interventions in BCT and AIT that have been studied concentrate on physical training program modifications. High running mileage has been identified as an injury risk factor, and an effective intervention for prevention or reduction of injuries in BCT is to decrease the amount of running mileage.[21] Targeted reductions in running mileage can reduce injury risk without significantly affecting aerobic fitness improvements.[18,161]

It is also important to conduct a fitness program with a large variety of physical training exercises that follow the principle of gradual and progressive exercise stress (overloading) to increase aerobic fitness, muscle strength and endurance, power, and neuromuscular conditioning (ie, proprioceptive skill development). Proper conditioning is as important for women as it is for men. Lower fitness levels are associated with higher injury rates in BCT.[4,6,9] Because of the risk factor, standards for fitness were established. Beginning in 1999, new female recruits reporting to BCT must be able to meet the minimum standard of three pushup repetitions, 17 situp repetitions, and a 10.5-minute or less 1-mile run before they even leave for BCT.[21] Increasing physical fitness in general will help prevent many overuse and traumatic injuries in military women. It should be done gradually, avoiding excessive running distances and frequency.

Progressive overload strengthening exercises of many of the lower extremity muscles (eg, the gastrocnemius, soleus, posterior tibialis, anterior tibialis, quadriceps, hamstrings, and gluteus medius) should be performed early and regularly in BCT and AIT. These exercises can help prevent overuse injuries by preparing these muscles for better adaptation to the demands placed on them. Specifically, the chronic use of eccentric contractions can lead to protective adaptations, such as increased muscle-tendon stiffness.[162]

As previously described, commanders should be strongly involved in injury prevention. They should maintain an emphasis on injury control education, recommend that trainees meet minimal physical fitness goals, and obtain new running shoes before starting BCT. Physical training and marching should be conducted in accordance with *Field Manual 21-20* ("Physical Fitness Training"), Army Regulation 350-1 ("Army Training and Education"), and *Field Manual 21-18* ("Foot Marches").

Running Shoes

"Shoes worn during physical training may be the most important equipment related to prevention."[1] Without appropriate supportive shoes specifically designed for running, women are at an increased risk for overuse injuries in the foot, ankle, shin, knee, hip, and back. There is no one running shoe that is perfect for everyone, but many experts have tips for purchasing an appropriate shoe. Running shoes should be comfortable, not old; have a strong and deep heel counter with an Achilles tendon relief; have a strong midsole (should not be able to bend the shoe in half); have a flexible toe box (should be able to flex under the ball of the foot); have a wide toe box; and have adequate cushioning in the heel. For a good fit, there should be a thumbnail's length between the big toe and the end of the shoe to allow for swelling and expansion during running. There should be plenty of room in the toe box, but no excess room in the heel. The age and condition of the shoes are also important. Shoes should only be worn for running (not daily outings or playing tennis) and changed after reaching approximately 500 miles or every 12 months (if running 3 miles/day at least 3 times/week).

Runners should buy running shoes that are appropriate for their foot types. For example, those runners with high arches (pes cavus or oversupination) will have "stiff feet" that prevent the normal pronation biomechanics that are important in shock absorption (Figure 13). This

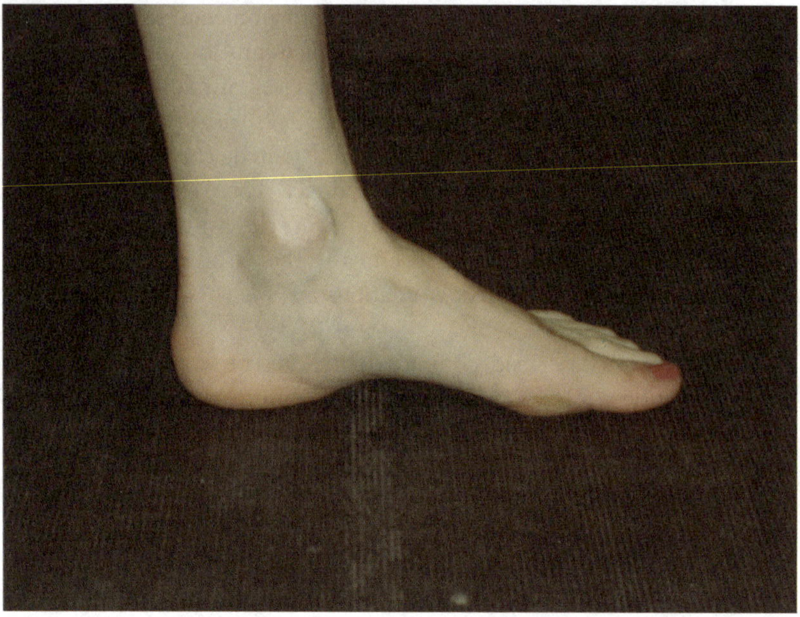

Figure 13. *Pes cavus.*

type of foot does not adapt well to the ground during running, and more stress and forces are translated to the bones, muscles, and tendons. These individuals should buy shoes with extra cushioning and a curved last. They should not buy shoes designed for motion control, because they need to be able to pronate during the stance phase of gait.

The majority of runners, however, need shoes with at least some motion control, because they have a predisposition toward pes planus or hyperpronation (Figure 14). This leads to excessive foot movement, and those individuals do better with a more stable and firm midsole and a straight last. The heavier the runner, the firmer the midsole should be. The heel counter should also be rigid to reduce excess rearfoot motion.

Those runners with normal arches can wear shoes designed for the neutral foot, typically called "stability" shoes. Shoe manufacturers recommend stability shoes for runners with neutral feet. During running, normal feet will hit the ground in supination at heel strike (80% of runners are heel strikers) and then roll into pronation as the foot continues to come in contact with the ground during the stance phase. The foot then supinates just prior to leaving the ground at toe off.

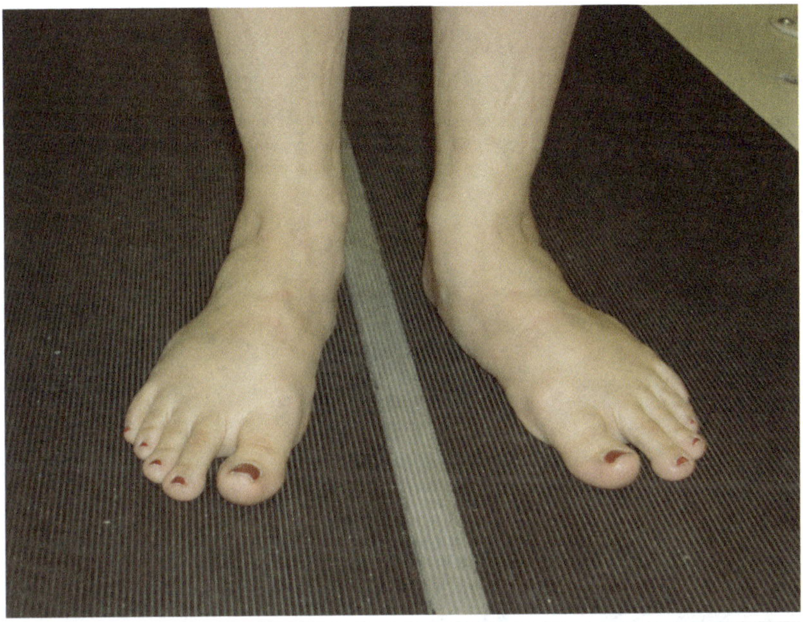

FIGURE 14. *Pes planus.*

The "wet test" is a basic, easy tool to teach runners about their foot types before they buy running shoes (Exhibit 1).[162a]

Multiple Intervention Programs

A few suggestions—such as training modifications, strengthening, conditioning, and running shoe selection—have been provided to aid in overall prevention of musculoskeletal injuries in military women. A combination of interventions will most likely be the best course of action to help reduce the incidence of injury. Combining suggested interventions may have a synergistic effect.[1] Researchers conducted a retrospective historical cohort study that reported injury rates following Australian recruit training. When compared with a preintervention group, injury rates for the postintervention group decreased 46% for men and 35% for women. The injury prevention program included cross-training (running in water), reduction of formation running distance and run test distance, interval training, controlling foot marching speed and load, and preventing running during foot marches.[163]

Use of multiple intervention programs have also been studied in US Army BCT and AIT, with success in reducing injuries without compro-

Exhibit 1. *Selecting the Right Running Shoe*

The "wet test" is an easy way to educate patients about foot type. Patients step on a dry surface with a wet foot. Then, they look at the shape of their footprint and compare it to these pictures. Once they determine their foot type, they will be able to buy the type of shoe they require.

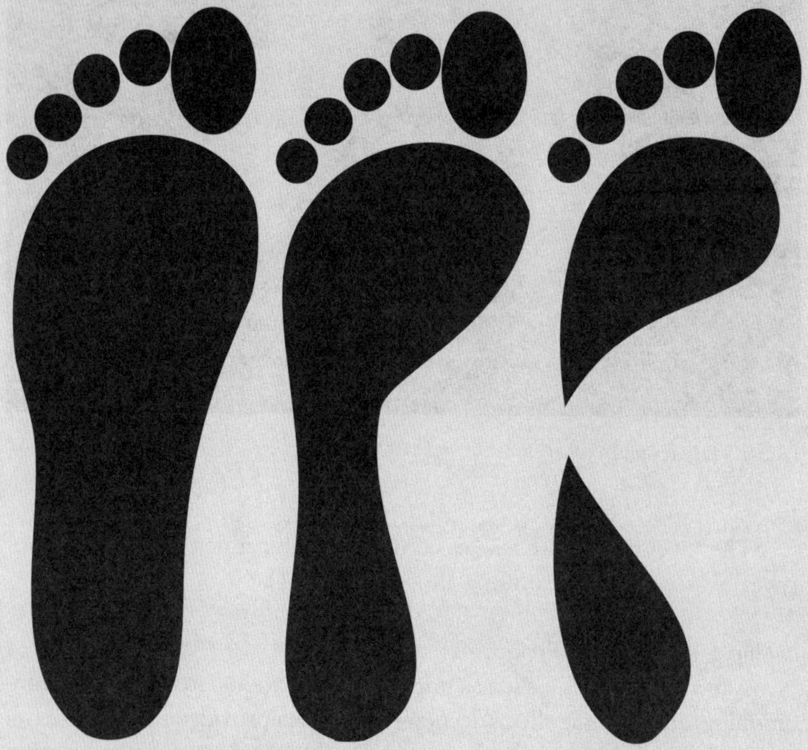

Flat Foot
(over pronator)
requires Motion Control Shoe

Normal Foot
requires Stability Shoe

"High Arch" Foot
(over supinator)
requires Cushioned Shoe

mising physical fitness.[7,8,164–166] Examples of the interventions from these studies include reducing running mileage, having no preexercise stretching, providing a wide variety of exercises using a progressive overload principle for strengthening, including interval training and movement drills, and educating the cadre on injury prevention. These interventional studies demonstrated that control group women ranged from 1.4 to 1.6 times more likely to be injured than the experimental group women who participated in the injury prevention program. The male control groups ranged from 1.5 to 1.6 times higher risk of injury compared with men in the experimental groups.

A new physical training program for all US Army BCT units was implemented in April 2004, and involves gradual and progressive overload, reduced running mileage, no preexercise stretch, and a greater variety of exercises. It is called the TRADOC Standardized Physical Training Program. Although it has not yet been fully analyzed, a preliminary analysis of injury rates (from 2003 to 2004) demonstrates an approximately 23% reduction across all BCT posts.[21] The TRADOC Standardized Physical Training Guides for BCT and AIT can be found online at http://www.benning.army.mil/usapfs/Dotrine/. Specific recommendations for physical training and injury prevention from this guide[1] include the following:

- Deemphasize distance running.
    - Use ability groups and interval training.
    - Do not run for physical training on consecutive days.
    - Treat foot marches more than 3 km as if they were running activities.
    - Strictly enforce heat injury prevention work/rest ratios.
    - Do not exceed 25 miles for total amount of running during BCT for trainees with average and below average fitness.
    - Use a standardized, gradual, systematic progression of running in BCT and AIT.
    - Rebuild fitness gradually for trainees who miss more than 1 week of physical training.
- Balance the program to prevent overtraining.
    - Do not overemphasize physical training as preparation for the fitness test.
    - Balance cardiovascular endurance with strength, mobility, and agility.
    - Count near-max or exhaustive military training as the equivalent of a heavy physical training session.
    - Consider the number of miles logged by units who must foot march great distances to and from training sites.
    - Perform low-intensity, task-specific, dynamic warm-up activities prior to more intense training in preference to stretching exercises.

  - Discourage formation running and cadence calls while running. (This is because people do not all run at the same pace and stride length. Women who are shorter end up taking longer strides, thus causing hip problems, such as stress reactions/fractures to the pubic rami.)
  - Eliminate remedial physical training programs that involve extra long training or more than one exercise session per day.
  - Allow adequate recovery; running and strenuous foot marching should not occur more than three (nonconsecutive) times/week.
  - Remember that the psychological and motivational aspects of overtraining can lead to attrition in BCT and AIT.
  - Do not use physical training as a punitive, corrective, or disciplinary tool.
- Ensure command responsibility for injuries, as well as for physical performance.
  - Note that commanders should assume responsibility for physical training outcomes.
  - Place more emphasis on the percentage of trainees passing the fitness test rather than the highest average unit score.
  - Use unit injury rates as a barometer of physical training program success or failure.
  - Emphasize that if average unit fitness scores are used, "zero" scores for those who cannot take the fitness test because of an injury profile should be included in the score.
  - Note that commanders should monitor injuries and performance, and report findings to higher headquarters.
  - Consider unit fitness test pass rates and injury rates when rating officers and NCOs.

General Treatment and Rehabilitation Strategies

Despite having interventions in place, injuries will still happen. If injury does occur, treatment should not be delayed and typically includes rest (physical profile/activity restriction), ice, compression, elevation (RICE), and nonsteroidal antiinflammatory drugs (NSAIDs) to reduce pain, swelling, and inflammation, and to lessen or stop the injury-producing cause.[1] Early diagnosis, "controlled rest" (cross-training), and optimal rehabilitation are keys to success in preventing further injury and/or chronic conditions. "Delaying evaluation and treatment may ultimately result in

TABLE 1. *Example of a Return to Running Progression*

	Walk	Jog	Repetitions	Total Time
Stage I	5 min	1 min	5 times	30 min
Stage II	4 min	2 min	5 times	30 min
Stage III	3 min	3 min	5 times	30 min
Stage IV	2 min	4 min	5 times	30 min
Stage V	Jog every other day with a goal of reaching 30 consecutive minutes			

a slower recovery or a worsening injury."[21] Physical therapy (PT) should be used early to aid in evaluation, education, manipulation, mobilization, stretching, and gradual strengthening exercises using the overload principle. The goal is to prevent chronic conditions and to help return service members to training or work as quickly and as safely as possible.

Returning an injured military woman to running is one of the goals that healthcare providers must assist with because it is an activity required of service members. One of the most important tools used to help a female service member return to running is an effective walk-to-jog program. After an injury, she should not be expected to begin at the same level of preinjury running distance, intensity, or frequency right away. Once the symptoms have subsided, she needs to follow a gradual running progression program. After she can walk at least 2 miles without pain or swelling, she can begin a walk-to-run program (Table 1).

The military woman should follow these six basic guidelines during this program: (1) if the jogging hurts, she must apply ice when finished and return to the previous stage the next day; (2) if there is no pain during or after running and she feels fine the next morning, she can proceed to the next stage; (3) if swelling develops in a joint or there is muscle/tendon pain that lasts longer than 72 hours, she has done too much and needs to decrease activity (duration and/or intensity) and increase rest between workouts; (4) she should increase the intensity of the jog before increasing the duration; (5) she should follow the 10% rule—only increase the weekly mileage by 10% of the previous week; and (6) she should use ice immediately after activity for 15 to 20 minutes early in this program. Command knowledge and support of these guidelines will facilitate recovery and return to running.

Specific Injury Prevention, Treatment, and Rehabilitation Techniques in the Military

"With gradual progressive physical overload training, stressed tissues can theoretically repair and recover from day to day, averting injury, pain, and loss of function."[21] However, women who are less fit will not be able to recover and may be overloaded to the point of pain and injury. In addition, training errors may exist. Specific guidelines for the prevention and treatment of common overuse and traumatic injuries suffered by military women are provided in the following sections.

Stress Fractures

Prevention. Improper training can cause bone tissue to reabsorb faster than it can regenerate, thus disrupting the ability of bone to repair itself in response to stress. One very effective way to prevent stress fractures is to decrease running mileage. There is an association between stress fracture incidence and running distance. Researchers from the US Naval Health Research Center (San Diego, CA) tracked Marine recruits during a 12-week Marine Corps boot camp and found that a 40% reduction in running distance was associated with a 53% reduction in stress fracture incidence without significantly compromising aerobic fitness.[167]

Running shoe age has a significant impact on risk of incurring a stress fracture in military recruits.[3] Those recruits wearing newer shoes were less likely to sustain a stress fracture. This is most likely associated with age-related loss of shoe cushioning and support.[1] Boots may also have an effect. Women recruits in Navy basic training who wore boots with greater shock absorbency had significantly fewer lower extremity injuries, fewer podiatric visits, and fewer severe stress fractures than those who wore standard Navy boots.[168] Another important prevention tool is to ensure that running shoes are appropriate for foot type. Women with high arches should wear shoes designed for cushioning. Women with pes planus/hyperpronation should wear shoes that provide motion control. (For more information about running shoes, see section on Overview of General Injury Prevention, Treatment, and Rehabilitation Techniques in the Military.)

Multiple interventions help reduce the incidence of injuries such as stress fractures. Female Australian recruits were studied retrospectively to report the incidence of pelvic stress fractures after the following five interventions: (1) reduce road march speed from 7.5 to 5 km/h without reducing the mileage, (2) march with their own stride length, (3) space out the marching and running formations to identify obstacles easier, (4) run on grass instead of roads when possible, and (5) reduce total running distance and substitute interval running occasionally for longer distance runs. The incidence of pelvic stress fractures in the intervention group was 0.6%, compared with 11.2% in the preintervention group.[169]

Treatment. Treatment must include diagnosing and correcting any predisposing or intrinsic factors, such as hormonal, nutritional, and biomechanical abnormalities. Stress fractures may be broadly categorized as either low-risk or high-risk injuries. Low-risk stress fractures typically heal without complication and can be diagnosed on the basis of a thorough history, physical examination, and radiographs. With high clinical suspicion, even with normal radiographs, a trial of rest and serial examination and radiographs is appropriate.[56] One to six weeks of limited weight-bearing with progression to full weight-bearing as symptoms allow may then be followed by a trial of low-impact activities, such as swimming or biking. Once the patient can perform low-impact activities for prolonged periods without pain, high-impact exercises may be initiated.[56]

Because the female athlete triad (defined as amenorrhea, disordered eating, and osteoporosis) is a strong risk factor for stress fractures and is a multifactorial problem, treatment should focus on all factors. Referral to a physician who is skilled in the evaluation and treatment of the female triad is recommended.[22] A nutritional and menstrual history is important in the treatment of stress reactions/fractures. Hormone replacement may be indicated in those with menstrual irregularities, such as amenorrhea (absence of menstruation) or oligomenorrhea (menstrual cycle greater than 36 days). Bone densitometry should be considered, especially if the service member suffers from multiple or recurring stress fractures. Psychological evaluation and therapy are recommended for those with eating disorders and chronic stress. Biomechanical factors should also be evaluated and corrected.

Hip Stress Fractures

Treatment of hip stress reactions/fractures will also depend on the area of the hip involved. According to a February 2006 draft of the "US Army Medical Command Hip Pain Clinical Management Guideline," any soldier in BCT, AIT, or an Officer Basic Course who presents for medical care with a complaint of hip pain and who demonstrates an antalgic gait, limited hip range of motion (less than 90 degrees flexion and 45 degrees abduction), or an inability to fully weight bear on the involved limb will receive baseline hip radiographs (anteroposterior hip and pelvis/frog leg lateral).

Femoral Neck. If the radiographs are positive for femoral neck stress fracture, the soldier should be immediately placed on crutches toe-touch weight-bearing, be given a profile (limited duty) for 4 weeks, and referred the same day to orthopaedics and within 72 hours to PT. If the radiograph is negative, the soldier needs to obtain a bone scan or MRI no later than 72 hours and be placed on profile. Analgesics or antiinflammatory medications should not be prescribed at this point. If the bone scan or MRI is positive for femoral neck stress fracture (tension or compression side), the soldier needs to be placed on 4-week profile, referred the same day to orthopaedics and within 72 hours to PT. As described previously, a patient with a tension-sided femoral neck fracture should be surgically treated with internal fixation.[56,76,80] After surgery, the patient should remain non–weight-bearing for 6 weeks, followed by an additional 6 weeks of partial weight-bearing.[56]

Pubic Ramus. If the radiographs are positive for pubic ramus stress fracture, the soldier should be placed on crutches weight-bearing to tolerance, given a profile for 4 weeks, and referred within 72 hours to PT. If the radiograph is negative, the soldier should obtain a bone scan or MRI no later than 72 hours, be placed on profile, and not prescribed analgesics or antiinflammatory medications at this point. If the bone scan or MRI is positive for pubic ramus stress fracture, the soldier should receive a profile, be referred 72 hours to PT, and prescribed NSAID for analgesia. Treatment involves cessation of aggravating activity until the pubic tenderness is gone, usually about 6 to 12 weeks.[76] Full activity should not commence until the patient has undergone a progressive pain-free exercise and stretching program.

Tibial Stress Fractures

The initial treatment for the less common stress fracture on the tension side (anterior cortex) of the tibia is a trial of rest for a minimum of 4 to 6 months, with or without immobilization.[56] If nonoperative management fails, operative management may be indicated, which should include excision and bone grafting[170] and intramedullary nail fixation.[56]

Other tibial stress fractures need to be treated with rest/activity modification. In fact, 93% of tibial stress fractures heal with conservative treatment.[171] Sometimes crutches (or a cane) are required until the individual can walk pain free. Casts are occasionally required in a BCT or AIT environment to further protect trainees from outside forces and influences. Running must be stopped until the bone is healed. The duration of rest and activity modification will depend on the degree of injury. PT is encouraged to emphasize strengthening and flexibility exercises for the lower extremities. Cross-training (eg, swimming or cycling) should be initiated early to maintain a conditioned state while allowing the bone to heal. Running may be resumed using the walk-to-jog progression described previously after the patient has been asymptomatic for at least 4 to 6 weeks following a stress fracture.

Metatarsal Stress Fractures

Metatarsal stress fractures, if treated appropriately, appear to heal uneventfully, and delayed union or nonunion is rare.[82] Treatment consists of rest from the predisposing conditions (eg, running and marching) for 4 to 6 weeks or until asymptomatic. A rigid-bottom surgical shoe or a stiff-soled shoe is helpful to protect from the stresses during the pushoff phase of gait.[22] Occasionally, a cast is required until symptoms decrease. When walking is pain free and at least 4 to 6 weeks have passed, the walk-to-jog progression program may begin (see section on Overview of General Injury Prevention, Treatment, and Rehabilitation Techniques in the Military).

First Metatarsal. Stress fractures of the first metatarsal are treated in a short leg cast. A patient may be made non–weight-bearing if the clinician feels that there is a possibility of the metatarsal deforming into an elevated position during healing.[82]

Fifth Metatarsal. As described previously, fifth metatarsal stress fractures have a propensity for delayed union or nonunion, and have a high risk of refracture after nonoperative treatment.[84] Patients with prodromal symptoms, but negative radiographs, may be treated with non–weight-bearing and a semirigid brace that unloads the fifth metatarsal.[172] If radiographs are positive, or symptoms have been present for more than 3 weeks, then the patient may be treated with non–weight-bearing cast immobilization for 6 weeks or intramedullary screw fixation.[56] The literature supports surgical treatment for patients with a diaphyseal stress fracture and radiographic evidence of delayed union or nonunion.[172a]

Shin Pain

Chronic Exertional Compartment Syndrome

Prevention. CECS may or may not be preventable. However, a gradual introduction to running and marching, a leg strengthening and stretching program, and/or orthotics for underlying biomechanical or structural problems are recommended.

Treatment. Rest, ice, NSAIDs, and orthotics may allow a gradual and safe return to presymptom activities. However, relief is variable and not always long term. If the prevention techniques do not help and symptoms continue with running and/or marching, the military service member may have to do the following: (*a*) cease or reduce running and/or marching with a temporary or permanent profile (limited duty), (*b*) be released from the military, or (*c*) undergo a surgical decompression (fasciotomy). Surgical decompression of the affected compartment has been reported to have at least a 90% probability of producing significant improvement.[173–175]

Postoperative Rehabilitation. After surgical decompression, heel-to-toe crutch weight-bearing as tolerated is allowed, but activities that can increase swelling should be avoided, and the leg should be elevated above the heart as much as possible for the first 24 to 48 hours. Crutches can be discontinued when there is normal gait. Gentle active range of motion, stretching, stationary bicycling, and walking may begin immediately and progress as tolerated. Formal strengthening exercises can be started by the end of the second week. Scar massage with cocoa butter or vitamin E oil should begin when the incision is well healed. A walk-to-jog program and gradual return to full activities may be initiated when there is full strength and flexibility, no pain or tenderness in the leg, and no pain with walking at least 1 mile (see section on Overview of General Injury Prevention, Treatment, and Rehabilitation Techniques in the Military).

Medial Tibial Stress Syndrome

Prevention. A gradual progression into running and marching should always be part of any military physical training program. Women with excessive foot pronation may be vulnerable to the development of medial tibial stress syndrome.[22] Heel cord stretching may help prevent this condition, and should be conducted both with the knee straight and then with the knee slightly bent. Off-the-shelf or custom-molded orthotics may be indicated for those who exhibit hyperpronation of the foot during marching and/or running to relieve stress on the medial soleus. As previously described, excessive pronation or excessive velocity of pronation may increase stress and strain to the medial soleus. Therefore, eccentric strengthening of the soleus can help prepare this muscle for these forces. These should be conducted over the edge of a step with the knee straight and then with the knee slightly bent to stretch the gastrocnemius and the soleus, respectively.

Treatment. If left untreated, chronic medial tibial stress syndrome can lead to periostitis from the chronic avulsion of the periosteum by the pull from the medial soleus.[22] Prompt evaluation and treatment are keys to success. Rest, ice, and NSAID should be considered as early treatment adjuncts. A study was performed at the US Naval Academy wherein midshipmen diagnosed with shin splints were treated with either rest, antiinflammatory medication, heel cord stretching, heel pads, or casting. The most effective treatment was rest alone.[95] However, orthotics may be indicated if there is excessive foot pronation. Those with medial tibial stress syndrome should cross-train with activities that are pain free (eg, water running or bicycling). An eccentric gastrocnemius and soleus stretching and strengthening program should begin when tolerated, and should be done with the knee straight and then again with the knee slightly bent. The strengthening program should also target the posterior tibialis and the toe flexors, and can include towel curls, marble pick-ups, and ankle inversion tubing exercises. When pain free with walking at least 2 miles, the patient can begin the walk-to-jog program using appropriate running shoes (see section on Overview of General Injury Prevention, Treatment, and Rehabilitation Techniques in the Military). A common error that can lead to recurrence is to return to running too early or too quickly.[22]

Knee Overuse

Patellofemoral Syndrome

Prevention. No one knows the best preventative technique(s) for PFS, because the cause(s) can be very complex and multifactorial. Six suggestions for prevention include the following: (1) strengthening the quadriceps and gluteus medius muscles; (2) avoiding training errors, such as a rapid increase in distance, speed, intensity, or frequency; (3) avoiding running on hilly terrain; (4) maintaining flexibility in the lower extremities; (5) using proper running shoes; and (6) wearing orthotics for abnormal foot mechanics.

Treatment. There are also many suggested treatment techniques for PFS without consensus on the best method. The management for anterior knee pain is primarily nonsurgical. Suggestions include, but are not limited to, the following: (*a*) relative rest (cross-training); (*b*) ice; (*c*) running at own pace and distance on level surfaces; (*d*) NSAIDs; (*e*) PT for education about proper running shoes and/or orthotics (stretching of hamstrings, quadriceps, and ITBs; strengthening of quadriceps and gluteus medius muscles; and/or patellar taping); (*f*) and use of knee sleeve with patellar relief or patellar tracking brace. Another type of brace, called a Protonics brace (Empi, St Paul, MN) is designed to help reduce contact pressure on painful knees. Biomechanical studies have shown that patellofemoral contact pressures are lowest between 0 degrees to 30 degrees of knee flexion. Therefore, closed-chain, short-arc knee extensions performed within a pain-free range of motion are the cornerstone of quadriceps strengthening and treatment for patellofemoral disorder.[76,98] When subluxation of the patella is present, McConnell taping may be used in addition to vastus medialis strengthening to alter glide, tilt, and rotation of the patella.[176] In addition, teaching the correct hip and knee form while descending stairs is important (Figures 15 and 16). It is important to strengthen the quadriceps and the gluteus medius.

Surgical indications for patellofemoral pain are narrow and include surgical realignment for patients with pain secondary to lateral subluxation and/or lateral tilt.[76] Surgical treatment includes direct repair of the medial patellofemoral ligament to its attachments to the adductor tubercle and vastus medialis obliqus, lateral release, or a distal patellar realignment procedure such as the Fulkerson or Elmslie-Trillat procedure.[98] Additionally, patellofemoral chondromalacia may be treated surgically if conservative, nonsurgical treatment fails.

In general, it is difficult in the military environment to manage PFS because of the uncertainty of the causes, persistence, exact diagnosis, and potential malingering. However, physical rest may be one of the most efficient means of relieving symptoms.[177] Once symptoms have subsided, quadriceps strengthening exercises are a mainstay of therapy and should be performed primarily via pain-free, closed-chain contractions, thus ensuring the knees stay behind the toes in proper form (Figures 17 and 18). High-resistance, open-chain knee extension exercises should be avoided because of the significant patellofemoral joint reaction forces. Therefore, it is recommended that leg extension machines be avoided and leg press machines be used instead.

Iliotibial Band Friction Syndrome

Prevention. Two suggested ITB tendinitis prevention techniques are to (1) strengthen hip stabilizers, especially the gluteus medius; and (2) keep the ITB flexible. A tight ITB contributes to excess friction placed on the band as it slides over the femoral condyle during flexion and extension of the knee. In addition, excessive downhill running should be avoided, and a heel lift may be indicated if there is a leg-length discrepancy.

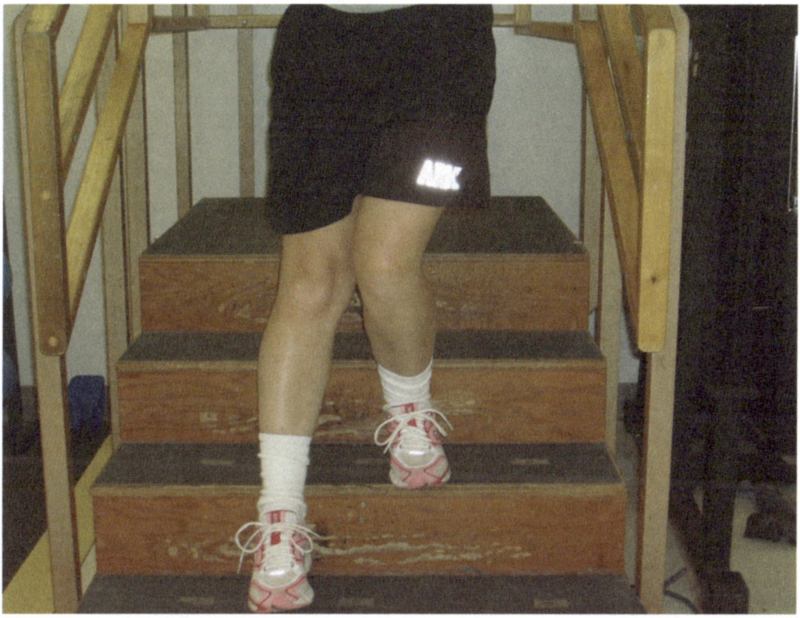

Figure 15. *Incorrect form: knee in valgus and hip in internal rotation.*

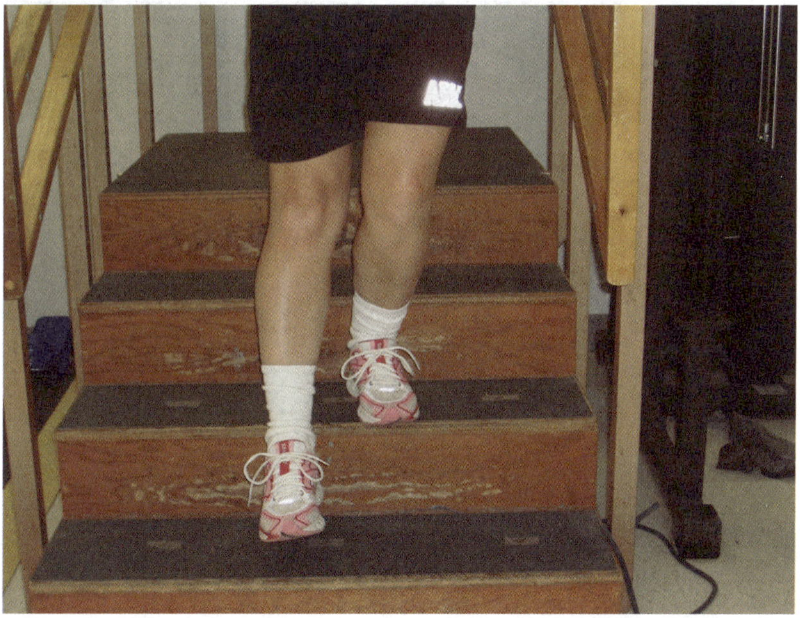

Figure 16. *Correct form: improved knee and hip control.*

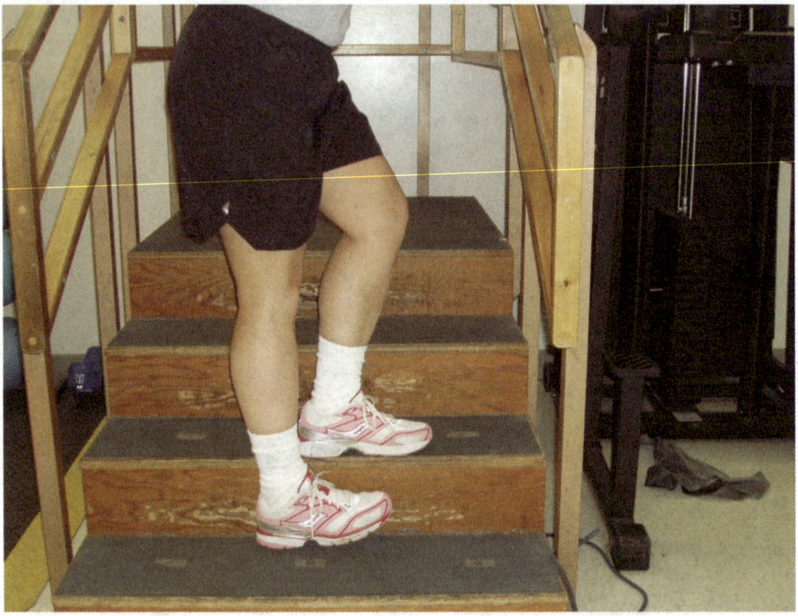

Figure 17. *Eccentric step downs (correct form).*

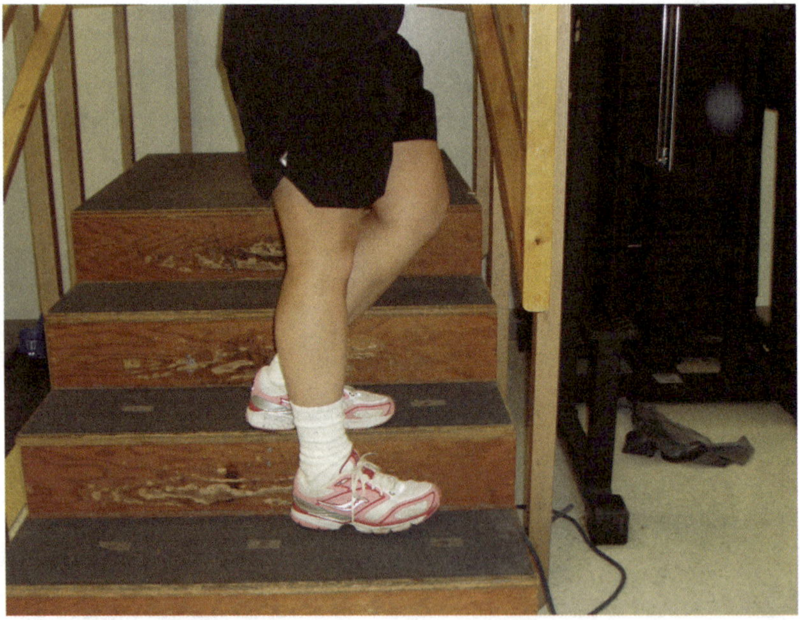

Figure 18. *Incorrect form (knee goes over toes).*

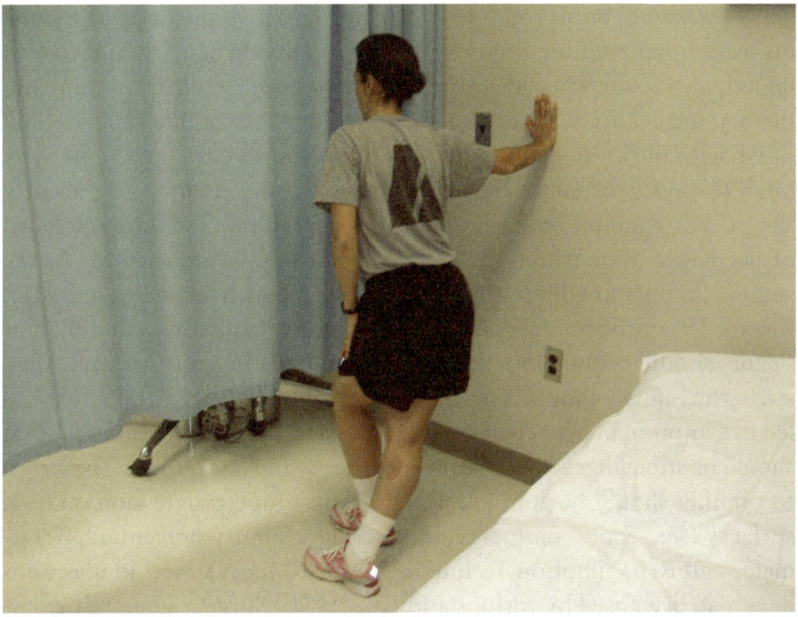

FIGURE 19. *Iliotibial band stretch.*

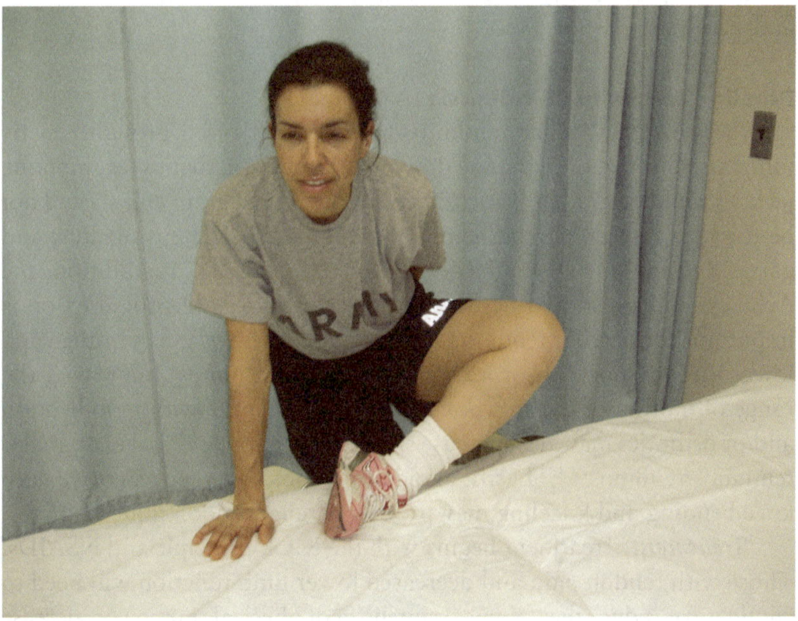

FIGURE 20. *Left piriformis stretch (lean forward at an angle).*

Treatment. Basic treatment principles of overuse injuries apply: decrease inflammation, limit overuse by modifying activity, and correct underlying factors when identified. This may require crutches, antiinflammatory medications, ice massage, stretching, a knee immobilizer and/or PT with ultrasound, and even transverse friction massage.[178] Activity modification is usually very effective and should be used in the early stages. For example, running (especially downhill) should be avoided or decreased until symptoms are resolved to minimize friction. As the acute inflammation diminishes, a stretching regimen focusing on the ITB (Figure 19), piriformis (Figure 20), hip flexors, hamstrings, and plantar flexors should begin. Once the patient can perform pain-free stretching, strength training should begin. Strengthening of the hip abductors has led to symptom improvement[179] (Figures 21 and 22). Other muscles that should be strengthened include the quadriceps and hamstrings. The return to running should be gradual, using a walk-to-jog progression on level surfaces (see section on Overview of General Injury Prevention, Treatment, and Rehabilitation Techniques in the Military). Steroid injections have also been used by various authors.[104,105,117] Surgery, generally excising the posterior portion of the ITB over the lateral femoral condyle, has been reported for recalcitrant cases.[103,113,114,117,180,181] Awareness of this syndrome, particularly in new recruits, is key to the correct diagnosis and treatment.

Patellar Tendinitis/Tendinosis

Prevention. Patellar tendinitis/tendinosis is often brought on by jumping, running, or marching downhill. Jumping activities (eg, jumping jacks) should be gradually introduced in the PT program and should not be used excessively. Strengthening the quadriceps, gluteus medius, and gastrocnemius-soleus should be part of the program. In addition, the Achilles tendons should be flexible. The range of ankle dorsiflexion is important for patellar tendinopathy, because it allows greater range and lower rate of shock absorption in landing from jumps. Improving the range of motion of the stiff or cavoid foot and ankle with mobilization and/or orthotics may also increase shock absorption and prevent patellar tendinitis/tendinosis.[182] Lastly, avoidance of prolonged or repetitive deep knee bending and kneeling may prevent patellar tendon injury.

Treatment. Treatment begins with the RICE principle and NSAIDs. Those with tendon pain and decreased lower limb function will need to eliminate or reduce the volume or intensity of physical training or increase the frequency of rest days. This can be accomplished with limited duty

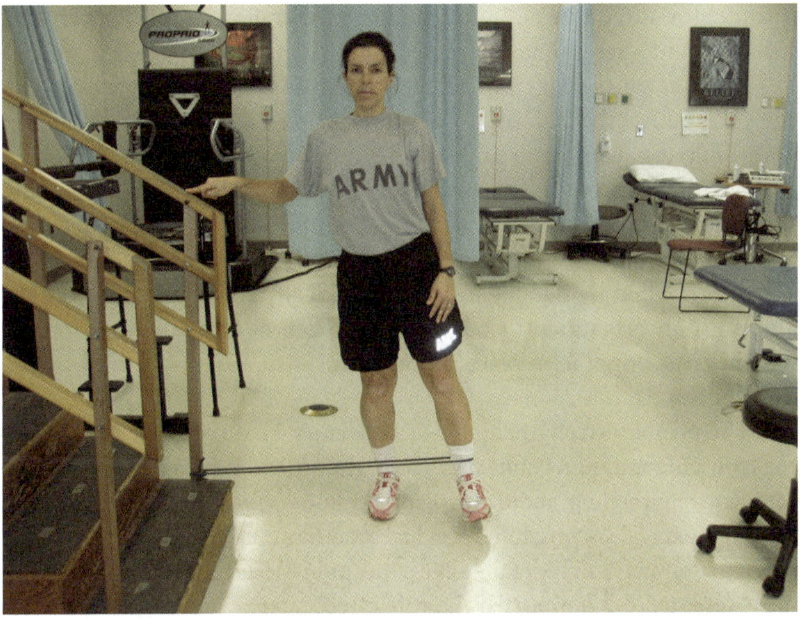

Figure 21. *Strengthening right hip stabilizers with well leg tubing perturbations.*

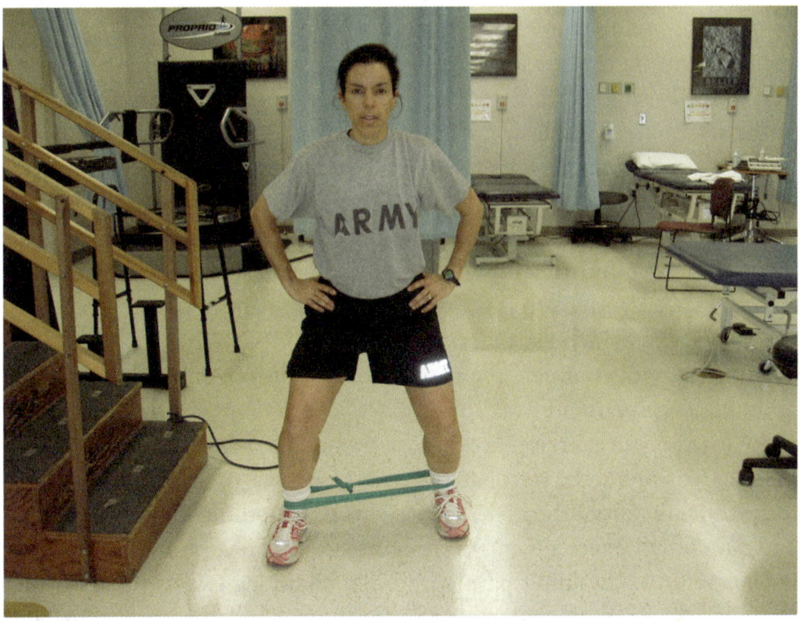

Figure 22. *Strengthening hip stabilizers with band "sumo walking."*

(profile) for 4 to 6 weeks. Jumping must be eliminated until the pain has subsided and strength has increased.

Patients who suffer from patellar tendinitis/tendinosis develop strategies to unload the painful knee. An example is landing from a jump with limited knee flexion and may be a result of weakness, pain, or a combination of both.[182] This usually leads to weakness in and around the symptomatic patellar tendon beginning with loss of power, followed by loss of endurance, and then loss of base strength.[182] Primary unloading is best identified by deficits in reduced capacity to squat. The patient's ability to absorb shock through the limb can be tested with hopping, whether it is hopping in place, hopping for distance, or performing the 6-meter hop.[183]

The primary strengthening exercises must be directed at the quadriceps musculotendinous unit specifically and should have a bias toward eccentric strengthening of the musculotendinous unit. The successful use of eccentric exercise in rehabilitation has been reported for many years. Curwin and Stanish[184] first detailed the concept in the early 1980s. Since then, most research has supported the role of eccentric exercise in improving painful tendinopathy.[185–190] Most rehabilitation must be undertaken as a single-leg exercise, since double-leg exercises allow compensation with the other leg.[182] The rehabilitation program should start with strengthening, progress to a power program, and finish with a sport/activity-specific load on the tendon. For example, quadriceps strength must be first gained using exercises such as leg press, decline squat, and lunges in the retropatellar pain-free range. Both legs can push the weight, and then the symptomatic leg can work eccentrically to lower/return the weight (Figures 23 and 24). There will most likely be some discomfort in the tendon because eccentric loading places the highest tensile loads through the tendon. The discomfort should not be disabling. Next, increasing speed with exercises, such as jumping rope and other low-level plyometrics, will increase tendon load. Sport/activity-specific loads, such as jumping, will complete the program. In addition, endurance must be an integral component of each level.[182]

Closed-chain exercises are important, but open-chain exercises—such as leg extensions in the retropatellar pain-free range of motion—are useful to specifically target the quadriceps. This is something that closed-chain and functional exercises may not be able to achieve; however, caution is warranted in avoiding pain with open-chain leg extensions. Weights provide excellent muscular loading with minimal tendon strain, and are therefore useful in the early stages of rehabilitation and throughout

FIGURE 23. *Concentric contraction (both legs).*

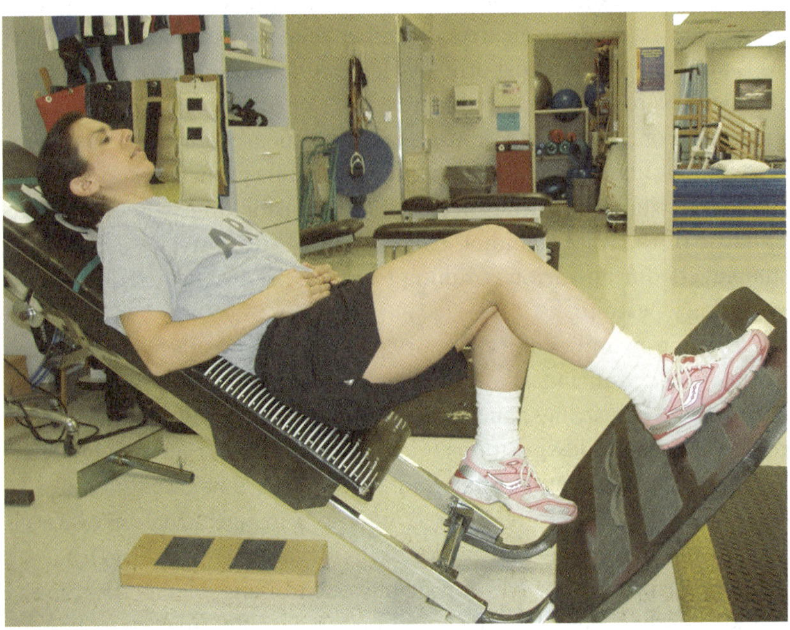

FIGURE 24. *Eccentric contraction (one leg).*

physical training.[182] In addition to the quadriceps, gastrocnemius, soleus, and gluteal functions are critical for adequate musculotendinous shock absorption. Close attention to all of these muscles during rehabilitation will maximize improvement in shock absorption.[182]

ACL Ruptures

Prevention. Prevention techniques and programs can be applied to help reduce the incidence of ACL ruptures. Intervention recommendations from the consensus statement (see section on Anterior Cruciate Ligament Ruptures) include the following six items: (1) conduct therapeutic exercises to reduce ACL and lower extremity injury rates, (2) modify body movement patterns and ground reaction forces via an intervention program, (3) conduct perturbation training to cause neuromuscular adaptations, (4) conduct plyometric training to increase quadriceps strength, (5) intervene with education, and (6) teach proper body positions and movement strategies.[128]

These recommendations can be applied to the military female population because the nature of the injuries is similar to that of the athletic population, wherein most of the ACL injuries occur. Sudden deceleration, tibial rotation, hyperextension, abrupt change in direction, and fixed foot are all possible causes of noncontact ACL injuries. Females in the military are all subject to these elements whether they are participating in BCT, AIT, physical training, airborne or air assault school, motor skills training, field exercises, sports and recreation, or even their occupations. For example, noncontact ACL injuries occur frequently during the deceleration phase of landing following a jump, or while preparing for a cutting maneuver while the knee is in near full extension and there is a valgus collapse of the knee. Landing from jumps off of obstacles and cutting maneuvers during various training exercises are frequent requirements of military personnel.

Specific examples of ACL injury prevention strategies in the military population can include the following four items: (1) strengthening the quadriceps, hamstrings, and gluteus medius early in BCT and AIT; (2) teaching proper body positions, movement strategies, and dynamic patterns, such as landing with hip and knees in more flexion and knees in less valgus (Figures 25 and 26); (3) incorporating perturbation exercises into physical training and motor skills development; and (4) having trainees perform plyometrics during physical training. Plyometric training (which is used to increase the speed or force of muscular contractions, thus providing explosiveness for a variety of sport-specific activities) is

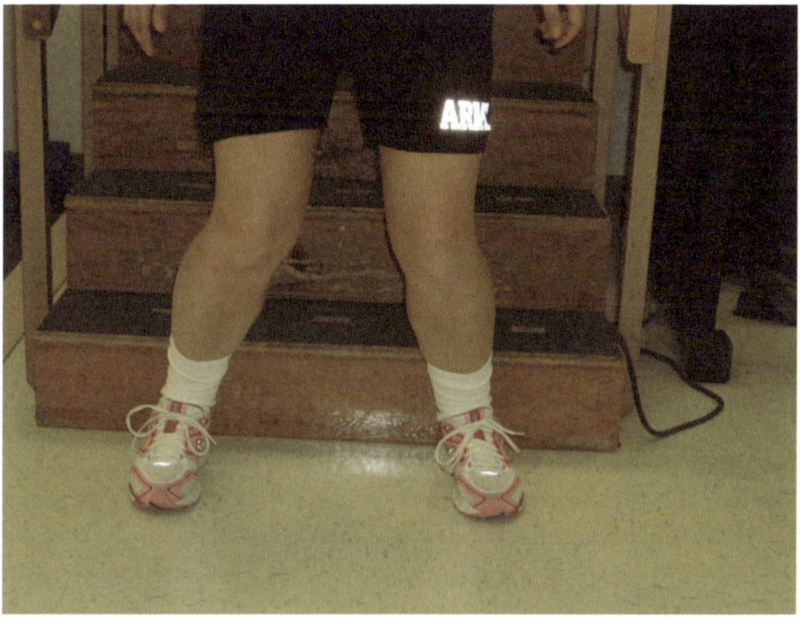

FIGURE 25. *Landing in too much knee valgus.*

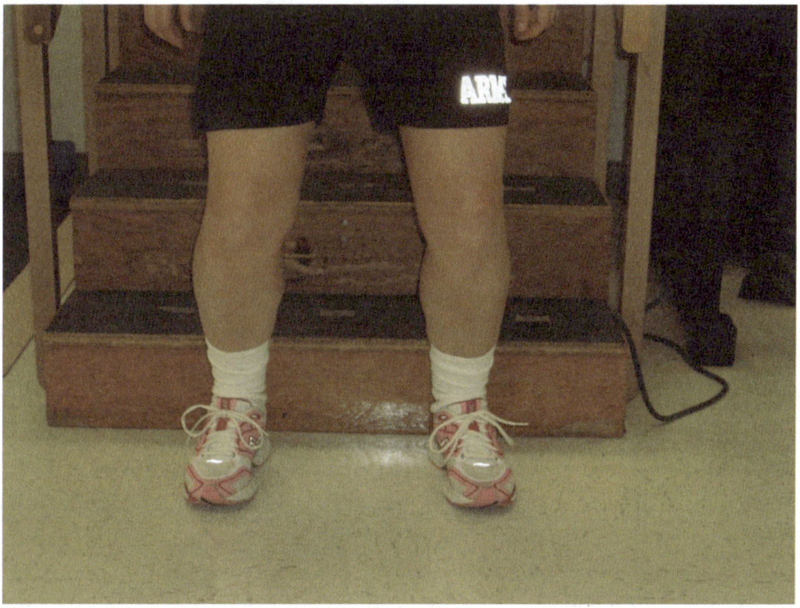

FIGURE 26. *Landing in less knee valgus (but not enough flexion).*

designed to decrease landing forces by teaching neuromuscular control of the lower limb during landing and to increase power. This training has led to increased jump heights, increased hamstring strength, decreased peak landing forces, and decreased knee abduction and adduction moments.[191] A plyometric training program can correct an imbalance of hamstring-to-quadriceps muscle strength and bring it to the level of males to restore hamstring-quadriceps muscle coactivation patterns.[191] This training could have a significant effect on knee stabilization and prevention of serious knee injury among military women. Hamstring muscles are important to help with stabilization of the knee joint because they function both as a joint compressor and a restraint to anterior motion of the tibia. These functions help decrease anterior shear forces and greatly reduce load on the primary restraint to anterior tibial motion: the ACL.

One other preventative technique would be to analyze the environment for hazards and make appropriate adjustments. One specific training hazard is rubber matting on obstacle courses. A report showed that the excessive coefficient of friction between rubber-soled boots and rubber matting increased the risk of ACL ruptures in Australian Army recruits most likely by fixing the foot to the ground, while the leg rotated in either the transverse or sagittal plane.[192]

Treatment. Those with an acute or chronic ACL-deficient knee may be able to function without gross instability and/or pain. However, if they participate in activities and sports that involve pivoting, cutting, or sudden starting and stopping, the chances of giving way and additional damage to the knee are greatly increased. Military service members routinely participate in such activities. If they suffer from acute or chronic ACL insufficiency and instability, they have the option of rehabilitation and trying to function in their current role with or without bracing, obtaining a temporary or permanent profile (limited duty), changing their occupation in the military, getting out of the military, and/or undergoing surgery.

Treatment for an acute ACL rupture should begin early in PT. NSAIDs and electrical stimulation will help with swelling and pain reduction. Ice, rest (a profile), crutches, and a knee brace are often provided. Patients are instructed to walk with a heel-to-toe gait to ensure full knee extension during the heel strike phase of ambulation. Backward walking is particularly helpful in regaining full extension. Obtaining full knee range of motion, especially extension, is critical for a successful conservative or surgical outcome and should begin within a day or two of injury. This can be aided by having the patient

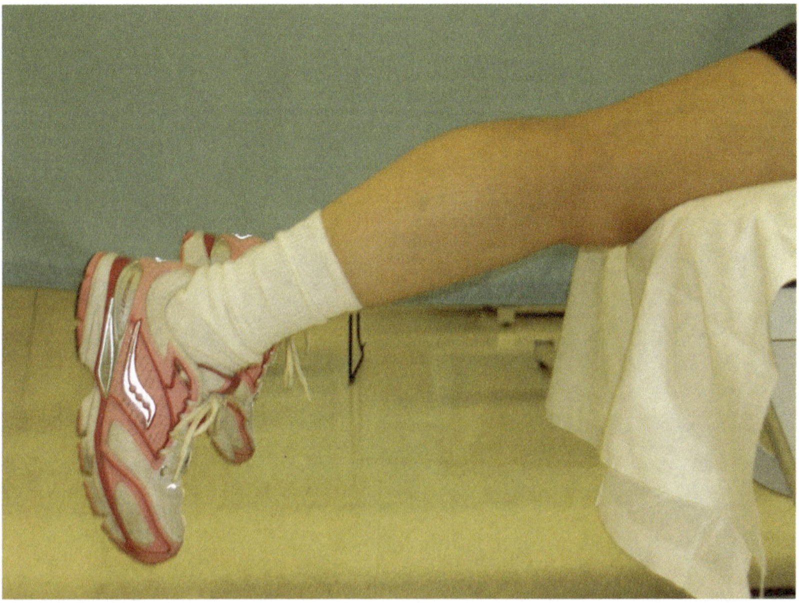

Figure 27. *Prone knee extensions (passive).*

lie prone with the knee over the edge of the table, letting gravity assist with knee extension (Figure 27). The patient can also perform standing terminal knee extensions (locking the knee back into extension with rubber tubing resistance) and prone terminal knee extensions with toes on the table and pushing the knee up toward the ceiling (Figures 28 and 29). Crutches can be discontinued when a normal heel-to-toe gait is achieved and quadriceps activity is restored. Stationary bicycle and closed kinetic chain quadriceps strengthening exercises can begin in the pain-free range. Hamstring, gastrocnemius, and soleus strengthening should also begin. Balance and proprioceptive exercises are necessary to aid in the functional stability of the knee. These activities should be initiated in a well-controlled environment and progressed as tolerated toward more challenging activities, such as standing on soft or moving surfaces, using perturbations, and incorporating trunk and upper extremity movements. Straight-line walk-to-jog progression may begin and progress as tolerated (see section on Overview of General Injury Prevention, Treatment, and Rehabilitation Techniques in the Military). Patients must be educated regarding the benefits and risks of whether or not to undergo an ACL reconstruction.

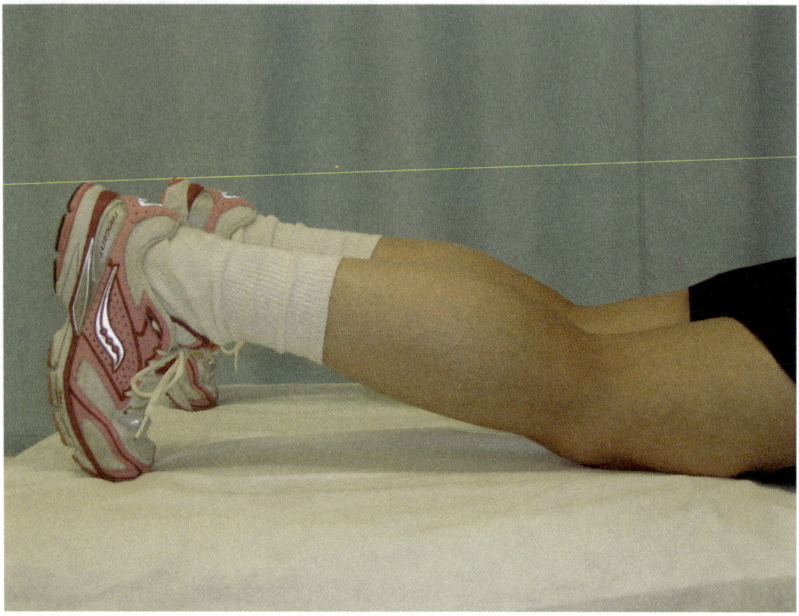

Figure 28. *Terminal knee extension—start.*

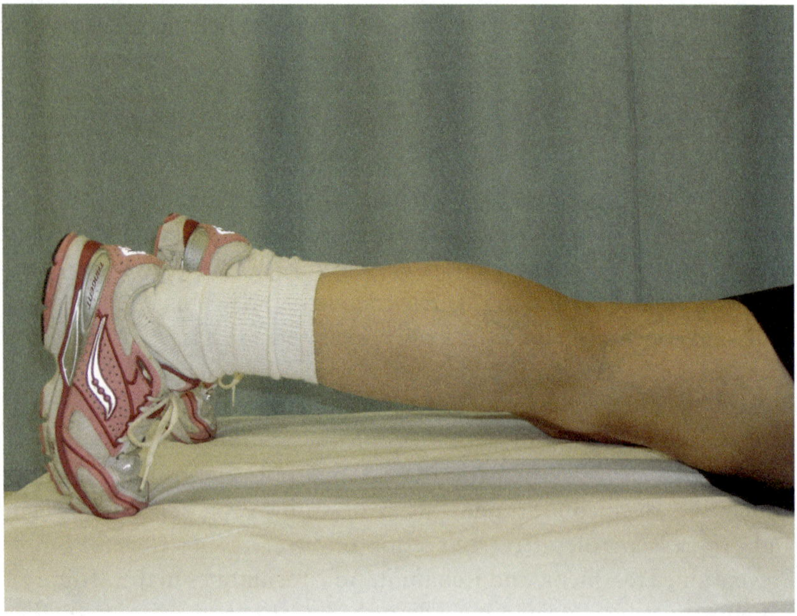

Figure 29. *Terminal knee extension—end.*

Ankle/Foot Conditions

Lateral Ankle Sprains

Prevention. Twenty-six percent to 41% of airborne operations injuries involve the ankle.[20,132–135] Exterior (outside-the-boot) ankle braces have helped reduce the incidence of airborne operations ankle injuries.[193,194] Others have reported that using prophylactic ankle braces can reduce the incidence of ankle injuries occurring from basketball.[195] In addition, running in formation, especially over uneven terrain, should be limited to avoid ankle sprains. The running and marching areas should be well lit and level.

Findings of a prospective study that examined intrinsic risk factors for inversion sprains in 159 young, physically active females suggest that effective prevention and conservative rehabilitation of ankle inversion sprains should consider coordination of postural control and improving ankle joint inversion position sense.[196] Balance, coordination, and timing drills should always be incorporated in physical training routines.

Treatment. If a sprain does occur, it should be evaluated and treated within a day or two of injury by a physical therapist and/or orthopaedist to help prevent chronic instability and pain. The Modified Ottawa Ankle Rules can be applied to determine the need for ordering radiographs following an acute ankle sprain.[197] Ankle fractures require immobilization, and grade III high-ankle sprains may need pinning, especially in the athletic and young active population. Other grade III sprains require bracing and extensive rehabilitation. Occasionally, surgical reconstruction of the torn ligament(s) is necessary if the joint remains unstable.

Most acute ankle sprains should be treated using the RICE principle, which includes controlled rehabilitation exercises, such as early motion and weight-bearing as tolerated with crutches. Athletic tape (open-weave technique), ace wrap, or an ankle brace can be used for immediate compression and protection. A U-shaped/horseshoe or circular pad can be placed around the lateral malleolus to provide focal compression (Figure 30). This will prevent excess fluid accumulation and further reduce edema. As described previously, crutches may be necessary if the patient demonstrates an antalgic gait, with weight-bearing status dependent on the severity of the injury. A heel-toe gait is also very important to help normalize function, prevent compensatory stresses at other joints, and help aid in venous return. Posterior splints should not be used following ankle sprains because they prevent ankle dorsiflexion and plantarflexion, which are necessary for a normal heel-toe gait and venous return. Ankle rehabilitation must also include progression to higher levels of strengthening, balance/proprioception, and functional skill training.

Achilles Tendinitis/Tendinosis

Prevention. Techniques to help in the prevention of Achilles tendinitis should include proper training overload progression for running, marching, and jumping, as well as eccentric strengthening of the gastrocnemius and soleus muscles. There is recent evidence that decreased plantar flexion strength and increased dorsiflexion range of motion are intrinsic risk factors for the development of an Achilles tendon overuse injury.[198] It is also important to match proper running shoes to foot type, such as pes planus (hyperpronation) or pes cavus. Running in other types of athletic footwear should be avoided at all costs. For example, running in basketball or court shoes can lead to Achilles injuries because the lack of arch support and the lower heel places more strain on the Achilles tendon with every step. This is especially true for those with increased dorsiflexion range of motion who are at increased risk already.

Treatment. Goals for the treatment of Achilles tendinopathy include the following: (*a*) reverse disease progression, (*b*) return the individual to previous level of activity unlimited by symptoms or other residual physical impairments, (*c*) prevent recurrence, and (*d*) help the individual manage his/her condition independently with education.[24] Treatment for acute Achilles tendinitis or strain should be prompt and include the RICE principle with NSAIDs. Resting is necessary, and a profile (limited duty)

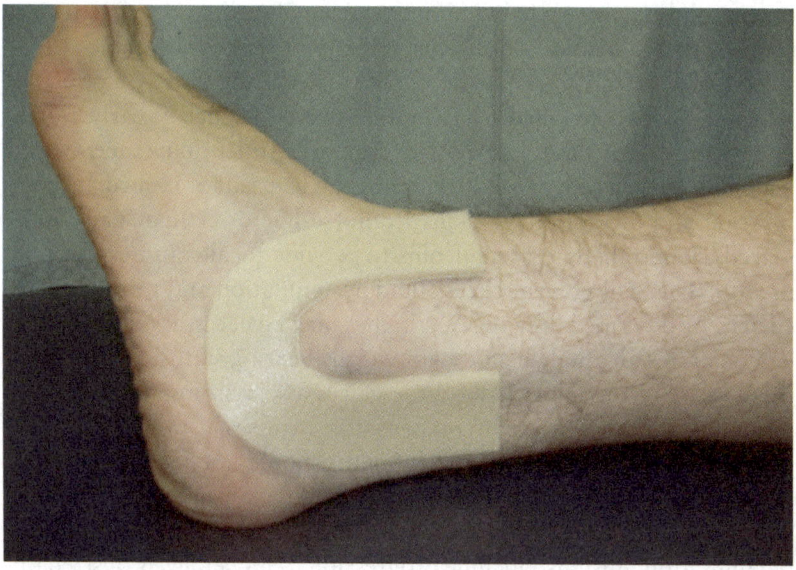

Figure 30. *Horseshoe pad.*

should be provided to prevent running, marching, and jumping for at least 4 weeks. Cross-training with biking or swimming will help maintain fitness levels while aiding in recovery. Ice and elevation help reduce pain, bleeding, and inflammation, especially during the first 72 hours. Compression, such as ace wraps or taping, may help reduce pain in the Achilles by unloading the tendon and restraining movement. Heel lifts in the shoes and boots also will help reduce the tension on the Achilles tendon by decreasing the amount of dorsiflexion during the terminal stance of walking. Orthotics should be considered if hyperpronation is present. Stretching and friction massage are also important following an injury to prevent inelastic scar tissue formation/cross-bridging. The stretching should include heel cord stretching, both with the knee straight and with the knee slightly flexed (Figures 31 and 32). Steroid injections should be avoided because of increased risk of tendon rupture.

Once the acute symptoms have decreased, an effective reloading program should begin and should include eccentric strengthening of the gastrocnemius and soleus. Eccentric reloading is effective following acute tendinitis and tendinosis. Research has demonstrated a high rate of return to preinjury activities in athletes who followed programs incorporating eccentric exercises.[185,186] The eccentric program should be progressed at slow speeds and sufficient loads to prevent continued or progressive tendinopathy. A recommended program is one outlined by Alfredson et al.[186] It consists of two types of eccentric exercises: (1) loading the calf muscles with the knee flexed and (2) loading the calf muscles with the knee straight (Figures 33 and 34). Individuals should start with 3 sets of 15 repetitions using just body weight, and only performing eccentric contractions—no concentric loading. This is achieved by using both legs to raise both heels, while using the impaired side only during the eccentric lowering phase in a slow, controlled fashion (Figures 35 and 36). The exercises should be performed two times per day, 7 days a week for 12 weeks. When only minor pain or discomfort is experienced, a backpack loaded with weight can be added. When walking 2 miles is pain free, a walk-to-jog progression program that is commonly used in military rehabilitation may begin (see section on Overview of General Injury Prevention, Treatment, and Rehabilitation Techniques in the Military).

According to Alfredson et al,[186] there is little place for surgery in the treatment of chronic Achilles tendinosis located in the avascular zone (2 to 6 cm proximal to the calcaneal insertion) in the tendon. In their study of the effect of heavy-load eccentric training in 15 athletes with chronic Achilles tendinosis, all participants had fast recovery in calf muscle function, and they returned to previous running activity symptom-free.[186]

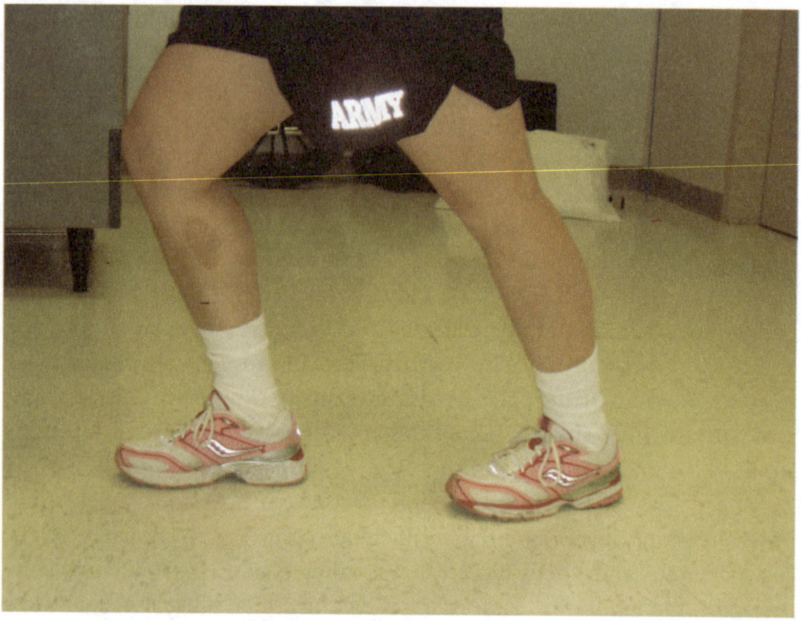

Figure 31. *Achilles stretching—knee straight.*

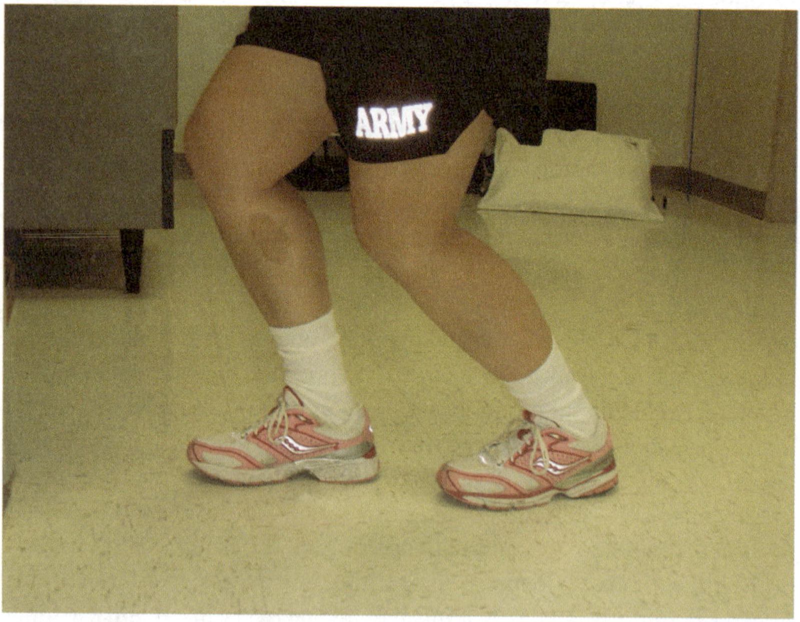

Figure 32. *Achilles stretching—knee flexed.*

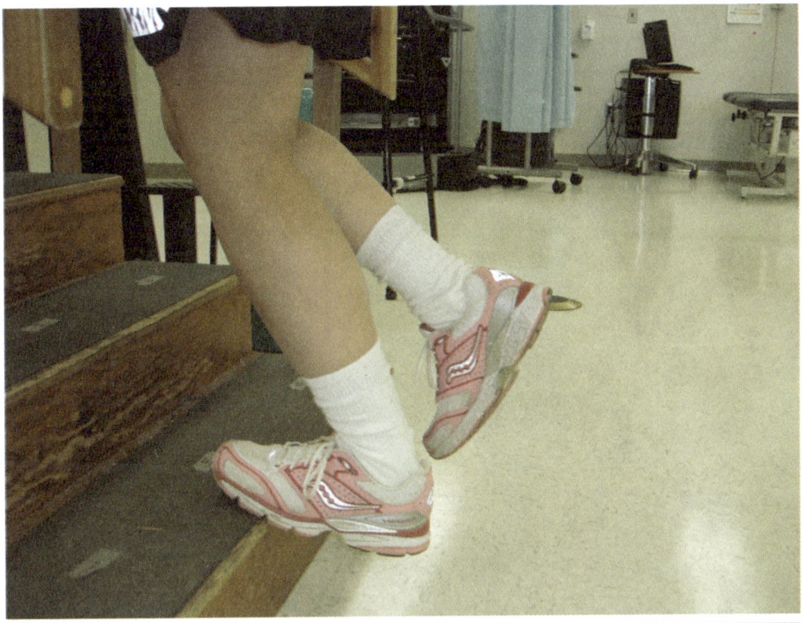

Figure 33. *Achilles strengthening—knee flexed.*

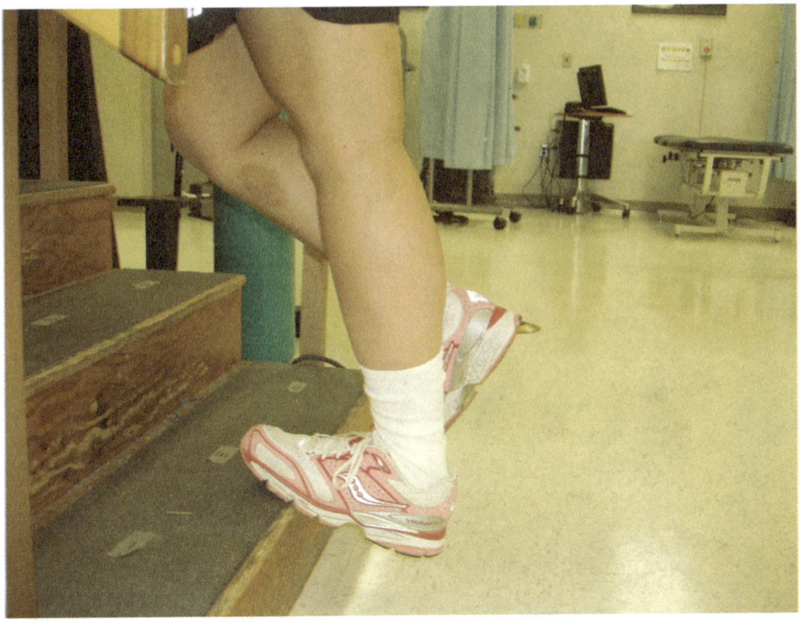

Figure 34. *Achilles strengthening—knee straight.*

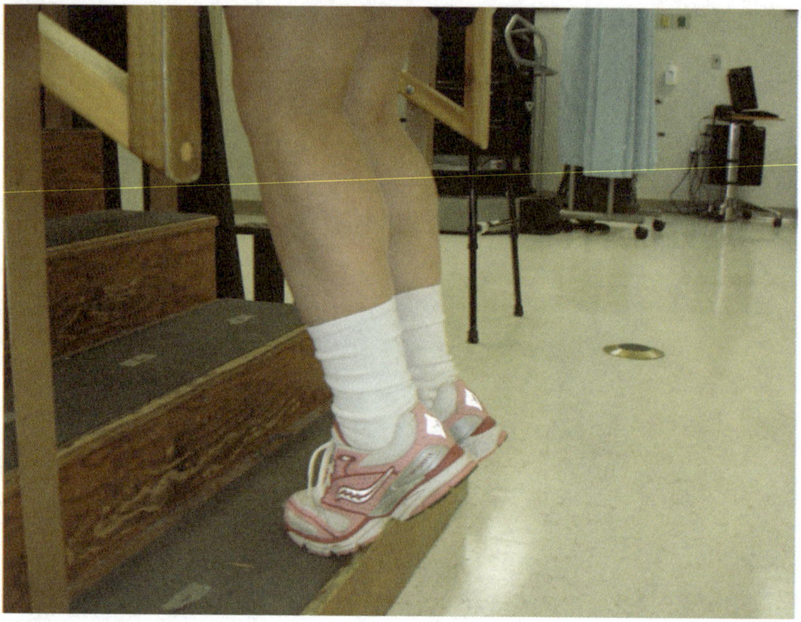

Figure 35. *Concentric contraction (both legs).*

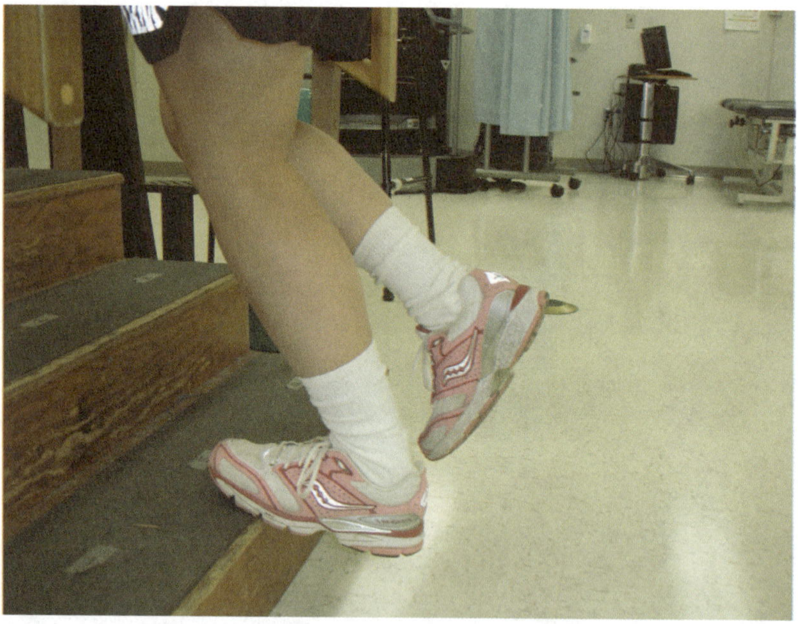

Figure 36. *Eccentric contraction (affected leg).*

Posterior Tibialis Tendinitis/Tendinosis

Prevention. Prevention of posterior tibialis tendinitis/tendinosis may be as simple as preventing training errors, stretching the Achilles, strengthening the posterior tibialis tendon, and ensuring use of proper running shoes. Women with excessive foot pronation are vulnerable to posterior tibial tendonitis, because this tendon has to provide eccentric control of foot pronation. Proper running shoes and/or orthotics can help limit the magnitude of foot pronation during closed-chain activities. In addition, if they have tight Achilles or limited talocrural dorsiflexion—especially if they wear high heels often—they are even more vulnerable because excessive foot pronation may be related to limited talocrural joint dorsiflexion range of motion.[24] Symptoms will be aggravated in those women who have occupations requiring long periods of standing, such as guarding post entrances (eg, military police).

Treatment. Treatment of posterior tibial tendinitis/tendinosis usually begins with the RICE principle and NSAIDs to help reduce clinical symptoms and prevent progression to chronic conditions and possible rupture. Patients should be educated about the condition to help prevent recurrence. Proper running shoes are a must, and orthotics may be indicated and usually include wedging or posting along the medial forefoot and hindfoot. Physical therapists can teach foot and ankle flexibility and strengthening exercises. These should include gastrocnemius and soleus stretching; eccentric strengthening of the posterior tibialis (ankle inversion) and toe flexors; and closed-chain strengthening of the gastrocnemius, soleus, and hip abductors in stance.

Chronic tendon disorders, such as posterior tibial tendinosis, often result from intensive repetitive activities (eg, running or marching). If higher-than-normal eccentric muscle forces are transmitted, the tendon may not be able to repair itself and may deteriorate to tendinosis. Ironically, the tendon appears to adapt favorably to physical stress, such as eccentric loading.[185,186] The tendons become stronger during the progression as fibroblast activity increases and collagen reaction accelerates. These high forces from eccentric loading induce a beneficial tissue remodeling response when exposed progressively and repeatedly. Cross-training with swimming, bicycling, etc, should also be part of the rehabilitation. Once symptoms are reduced and walking at least 2 miles is pain free, the walk-to-jog progression program can begin (see the section on Overview of General Injury Prevention, Treatment, and Rehabilitation Techniques in the Military).

Plantar Fasciitis

Prevention. Limiting the wear of military boots to those training activities that really require boots may be beneficial in BCT and AIT.[21] Other prevention techniques include proper running shoe selection, avoiding training errors, and orthotics.

Treatment. In BCT, common treatments for plantar fasciitis include ice massage, NSAIDs, Achilles stretching, correcting training errors, heel cups, and orthotics.[21] The rationale for using orthotics is to decrease the stress on the medial tuberosity and plantar fascia.[22] For those with pes planus, an orthotic device should be considered to correct the biomechanical deformity and to increase the support of the foot during the stance phase.[22] Athletic tape may also be indicated with this foot type to relieve the pressure on the origin of the plantar aponeurosis during propulsion by holding the forefoot in adduction and the heel in varus.[22] For those with pes cavus, a soft orthotic may be used to aid in shock absorption and to increase the area of contact.[22]

PT can help with the trial of modalities (ultrasound/phonophoresis): teaching specific foot strengthening exercises (eg, towel curls), plantar fascia stretching, and rolling the foot over a golf ball to release tension and break up adhesions or nodules; evaluating foot biomechanics and recommending/making specific orthotics; taping; and/or placing service members on limited activities (profile).

Acetabular Labral Tears

Prevention. Prevention of acetabular labral tears is difficult because up to 74.1% of labral tears are not associated with a specific event or cause.[139,140] However, it is believed that repetitive microtrauma is the cause of these insidious labral lesions,[139,199] wherein the pain gradually develops and increases in intensity over time. Hip stress fractures need to be ruled out. Physical training should be modified if a female begins experiencing groin and/or lateral hip pain, clicking, catching, locking, or giving way. If running causes pain, an alternate cardiovascular exercise can be used for cross-training. Specific motions to avoid are torsional or twisting movements, hip hyperabduction, hyperextension, and hyperextension with lateral rotation.[141] Straight-leg raising exercises should also be avoided if painful.

Treatment. Conservative treatment for hip labral tears includes PT, limiting/avoiding painful weight-bearing activities, cross-training with pain-free activities (eg, biking and swimming), and NSAIDs. The specific PT intervention for those with acetabular labral tears is not yet defined in the literature.[141] However, from clinical experience, PT can be beneficial for some patients and can also be effective for preoperative conditioning/strengthening if surgery is indicated. Specific recommendations include stretching the iliopsoas muscle to decrease the amount of traction on the labrum, stretching the hip joint capsule if motion is limited, performing core stabilization exercises targeting the transversus abdominis and multifidus (MF) muscles, and strengthening the surrounding hip musculature to stabilize and protect the hip joint. Performing well-leg tubing exercises in all directions and doing the "sumo walk" with rubber bands around both ankles are effective methods of strengthening the hip stabilizers, such as the gluteus medius (see Figures 21 and 22). They should be performed for time (endurance strengthening).

If nonsurgical treatment does not decrease or eliminate the symptoms, surgical intervention may be indicated. Surgical treatment has resulted in short-term improvement, but the long-term results are still unknown.[141] Repair and/or debridement of the torn tissue are typical surgical management strategies. Excision or debridement of the damaged labral tissue by arthroscopy is the most common procedure.[141] The PT postsurgical protocol used at Walter Reed Army Medical Center (Washington, DC) is provided in the Appendix.

Back Conditions

Prevention. Because military women have increased risk of low back pain—compared with men in specific occupations (eg, multichannel transmission systems operator, single-channel radio operator, wheeled-vehicle mechanic, signal intelligence analyst, voice interceptor, and practical nurse[144,145])—there may be different physical and psychosocial stressors involved. For example, there may be higher pressure to perform successfully or greater role conflict working in a traditionally male profession.[144] Therefore, there is a need for investigation, identification, and management of potential causes for gender differences. Specifically, supervisors must analyze high-risk jobs for psychosocial and ergonomic hazards and reduce their impact on pain and fatigue.[144,145] Ergonomic evaluation and education, safe lifting education, and core stabilization exercises are some examples of low back injury prevention techniques that are used in the military population.

Commanders and supervisors should use assistance from local PT or occupational therapy clinics, occupational health, or installation safety offices. The training should include teaching and emphasizing proper ergonomics, body mechanics, and the importance of being physically fit. An excellent source for supervisors is the US Department of Labor Occupational Safety and Health Administration (OSHA) Website (http://www.osha.gov).

Computer Workstation Ergonomics. Supervisors need to identify and minimize high forces on the spine and highlight the value of safe working postures, safe working conditions, and furniture design.Because of higher technology, many military personnel sit in front of computers throughout the day, which can lead to low back pain. OSHA provides thorough education about the best way to set up a computer workstation. (For more helpful tips, also see http://www.osha.gov/SLTC/etools/computerworkstations/positions.html.)

"To understand the best way to set up a computer workstation, it is helpful to understand the concept of neutral body positioning. This is a comfortable working posture in which your joints are naturally aligned. Working with the body in a neutral position reduces stress and strain on the muscles, tendons, and skeletal system, and reduces your risk of developing a musculoskeletal disorder." The following points are important considerations when attempting to maintain neutral body postures while working at the computer workstation:

- Hands, wrists, and forearms are straight, in-line, and roughly parallel to the floor.
- Head is level, or bent slightly forward, forward facing, and balanced. Generally, it is in-line with the torso.
- Shoulders are relaxed, and upper arms hang normally at the side of the body.
- Elbows stay in close to the body and are bent between 90 degrees and 120 degrees.
- Feet are fully supported by floor or footrest.
- Back is fully supported with appropriate lumbar support when sitting vertical or leaning back slightly.
- Thighs and hips are supported by a well-padded seat and are generally parallel to the floor.
- Knees are about the same height as the hips, with the feet slightly forward.

Regardless of how good the working posture is, working in the same posture or sitting still for prolonged periods is not healthy. The working position should be changed frequently throughout the day in the following ways:

- Make small adjustments to the chair or backrest.
- Stretch fingers, hands, arms, and torso.
- Stand up and walk around for a few minutes periodically.

Safe Lifting. A second important method of preventing low back pain and injury is to use safe lifting techniques (Figure 37). OSHA provides the following lifting tips, which also can be found at http://www.osha.gov/SLTC/youth/restaurant/poster_lifting.html.

DO

- Keep your head up and your back straight and bend at your hips.
- Bring the load as close to you as possible before lifting.
- Lift with your legs, not your back.
- Shift your feet to turn.
- Keep the load directly in front of your body.
- Try to perform lifts at waist height, with your elbows in close to your body.

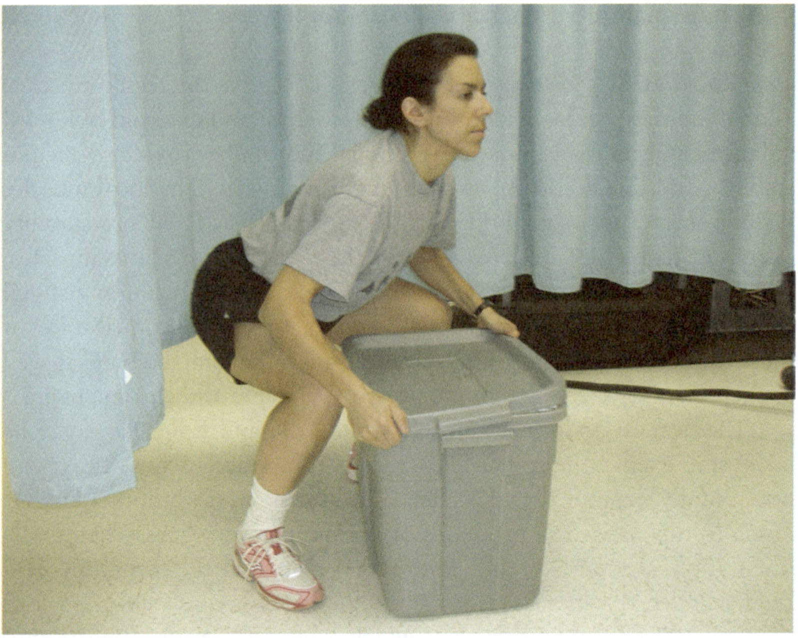

FIGURE 37. *Safe lifting technique.*

- Limit amount lifted by hand. Use mechanical lifts or get help.
- Stay fit to help avoid injury.

DON'T

- Lift heavy loads (>35 pounds). Get help.
- Reach across something to lift a load.
- Lift bulky or uneven loads.
- Reach to the side or lift while twisting.

Core Stabilization Exercises. Another important method in the prevention of low back pain is to improve the mechanical support of the spinal joints through lumbopelvic stabilization exercises (ie, core strengthening or stabilization).[200] Core stabilization has been used in athletic and therapeutic settings to enhance neuromuscular pathways, strength, proprioception, and balance, and to aid in coordination of synergistic and stabilizer muscles. Most importantly, strengthening core stabilizers has been shown to prevent low back pain.[201] Therefore, the goals of core stabilization are to control, prevent, or eliminate low back pain; increase patient education and kinesthetic awareness; increase strength, flexibility, coordination, balance, and endurance; and develop strong trunk musculature to enhance upper and/or lower extremity functional activities.

Global and local (deep core) muscles comprise the trunk muscles. Global muscles (rectus abdominus, and internal and external obliques) balance external loads, and local muscles internally provide segmental stability and direct control of the lumbar segments. The local muscles include the transversus abdominis (TrA) and the MF.[202] Core stability exercises center on the treatment of these deep core trunk muscles, which have been shown to be the most significant in sustaining spinal stability.[203] Recent works have focused on the functional contribution of these trunk muscles to postural stabilization of the lumbar spine, as well as their respective changes in the presence of acute and chronic pain.

The deepest abdominal muscle is the TrA, and it is believed to be a key stabilizer of the spine. The TrA is a thin muscle whose fibers run horizontally around the abdomen and attach via the thoracolumbar fascia to the transverse processes of the lumbar vertebrae. The muscle orientation is hoop-like, and contraction results in the formation of a rigid cylinder; this increases intraabdominal pressure, as well as tensions the thoracolumbar fascia, both of which increase stiffness in the lumbar spine.[202,204] This helps limit translational and rotational movements in the lumbar spine.[205] It also activates in an anticipatory manner to stabilize the spine before limb segment movement.[206–209]

Prevention efforts should target not only the TrA, but also the MF, pelvic floor, and diaphragm muscles both in motor learning principles and muscle strength endurance. These muscles work together in a cylinder-like effect to support the lumbopelvic region.[210] The MF is the largest and most medial of the lumbar muscles.[211] Its functions are to reduce anterior shear force, stabilize the trunk, and act as a primary trunk extensor. This muscle co-contracts with the TrA via the thoracolumbar fascia. Core stabilization focuses on whole-body equilibrium, lumbopelvic orientation, and intervertebral control via strengthening the TrA and MF muscles.[212]

Co-contraction of these deep muscles for core stabilization is a difficult task for anyone to learn. Education and training are keys to success. Pictures of the muscles and explanations of the exercises should be provided prior to practicing. The basic principles for this technique include (*a*) placing the pelvis in the most pain-free position (neutral spine) and then (*b*) drawing in the lower abdominal wall without moving the spine or pelvis.[210] This is to help isolate the TrA and MF from

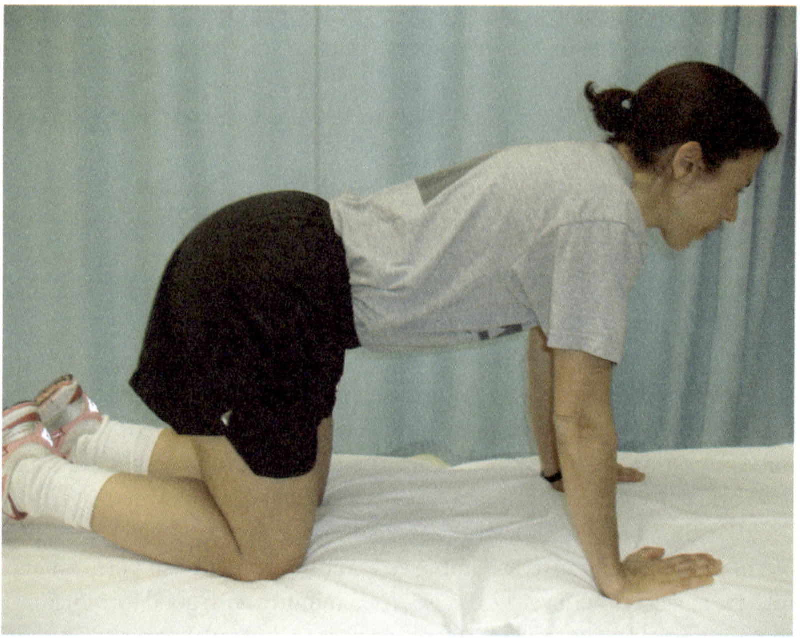

FIGURE 38. *Quadruped position to learn abdominal drawing-in maneuver.*

the global muscles. Helpful verbal cues include the following: "Take a relaxed breath in and out and then, without breathing in, slowly draw your lower stomach up and in towards your spine" and "continue normal breathing while holding it in" (Figure 38). Usually, this is first attempted in the hooklying position, but may need to be taught in the quadruped position using gravity for stretch reflex assistance. It is also important to explain that these deep muscles are comprised of type 1 muscle fibers, and endurance is the key to success.[210]

Real-time ultrasound imaging can be used to provide further biofeedback and to record outcome measures of the resting and contracted size of the TrA and/or the MF (Figures 39 and 40). Real-time ultrasound is a reliable noninvasive technique to measure thickness of these muscles. Data from the study by Teyhen et al[213] provide construct validity for the idea that the lower abdominal drawing-in maneuver results in preferential activation of the TrA in patients with low back pain. Results from this study provide further support for the use of real-time ultrasound as a foundational component for lumbar stabilization programs.

Once the basics of neutral spine and proper muscle contraction are learned and recorded, level 1 core stabilization exercises may begin. Participants must be reminded to maintain a neutral spine, draw in the lower abdomen, and gradually increase time spent performing the exercise. Examples of level 1 core exercises include the following:

- *Hooklying*: alternating hands to knees (Figure 41), bridging, and leg circles.
- *Prone*: one arm, one leg, or alternate arm and leg lifts.
- *Sitting*: on disc (Figure 42), arms overhead, one leg out, or alternate hands to knee.
- *Quadruped*: one arm, one leg (Figure 43), or alternate arm and leg lifts.

Advancement occurs by progressing from stable surfaces to unstable surfaces (ball, disc); large, simple movements to smaller, more complex movements; one plane of movement to multiple/combined planes; short lever arm to longer lever arm; no weights to weights; and slow to fast speed. Progression of core stabilization exercises occurs when level 1 exercises are mastered; endurance has improved; and the patient can perform and tolerate at least three level 1 exercises for 3 minutes with good form/neutral spine, minimal cueing, and loaded postures are tolerated. Examples of more challenging core stabilization exercises include the following:

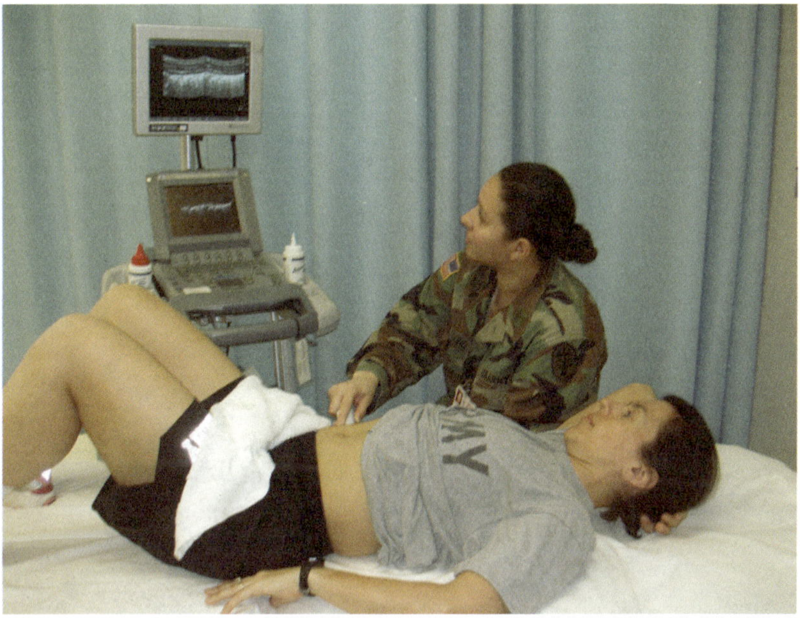

FIGURE 39. *Real-time ultrasound used for biofeedback.*

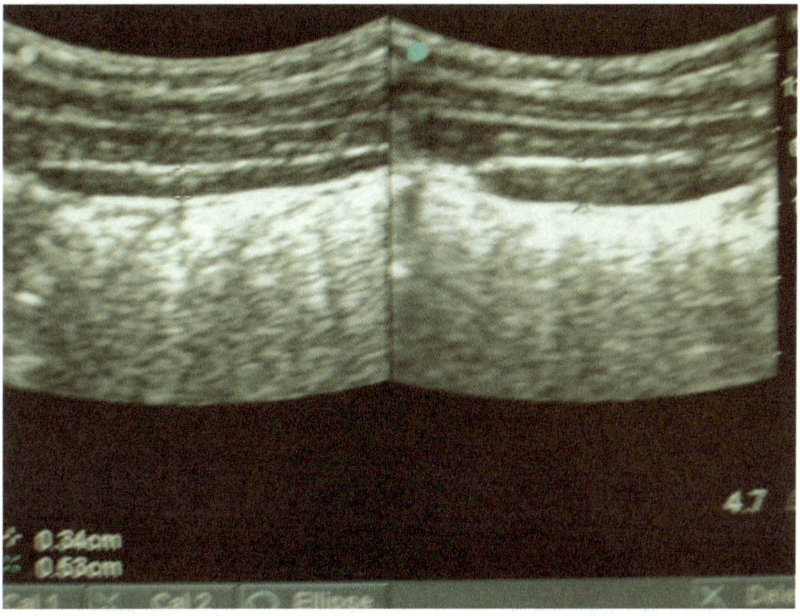

FIGURE 40. *Transverse abdominis relaxed (left) and contracted (right).*

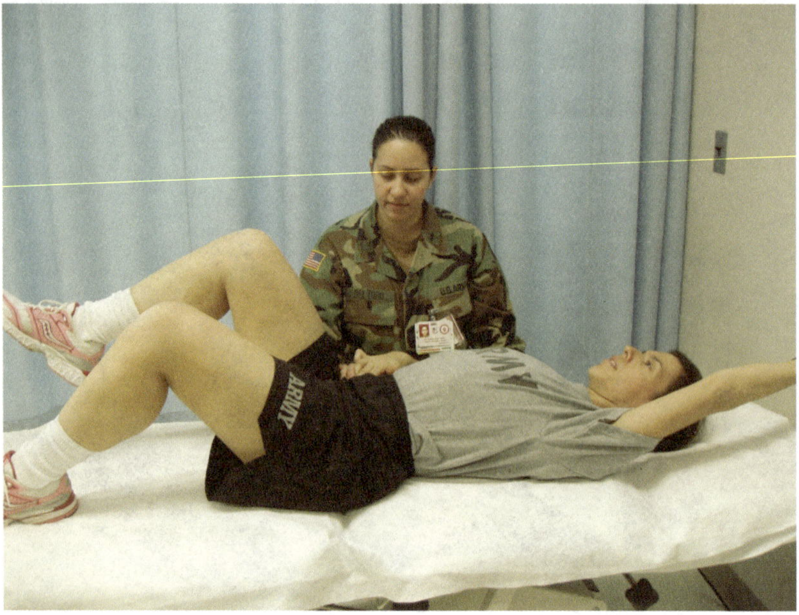

FIGURE 41. *Supine alternating arms and legs.*

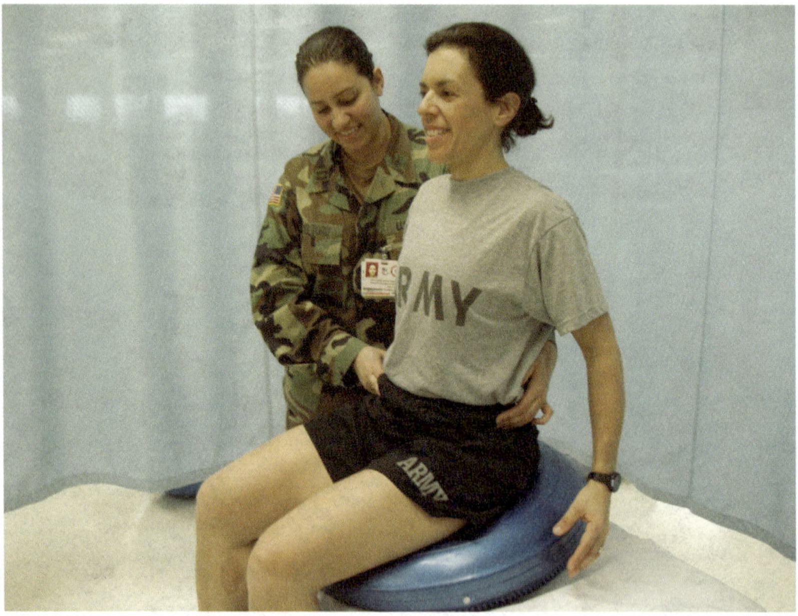

FIGURE 42. *Sitting on disc.*

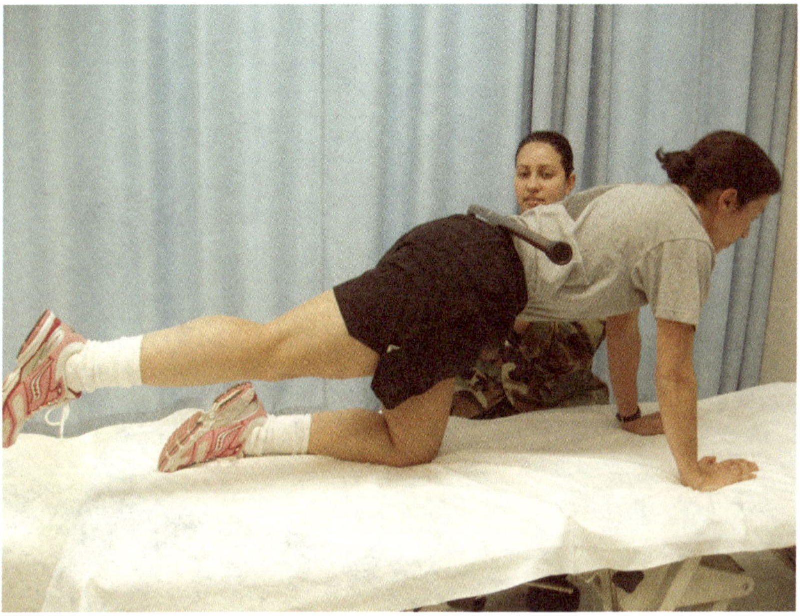

Figure 43. *Quadruped leg extension with cane on back.*

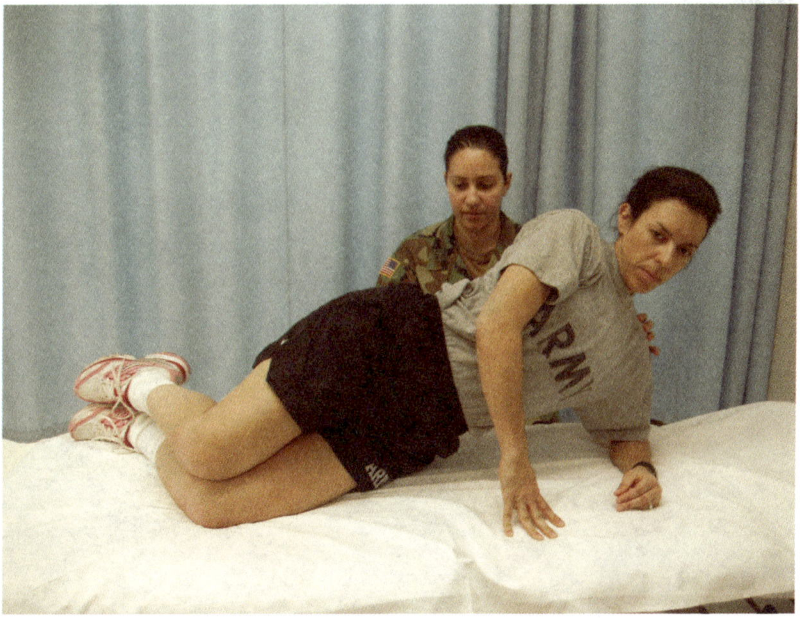

Figure 44. *Side plank.*

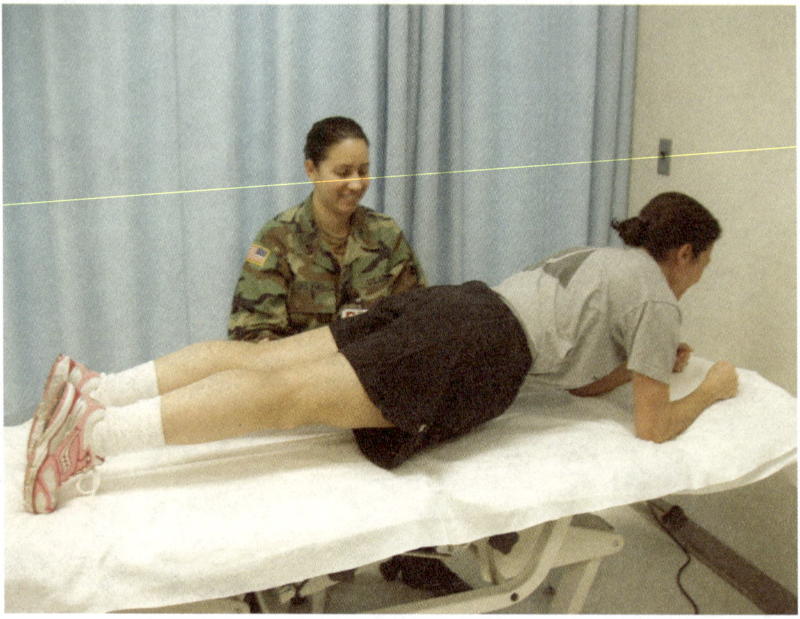

Figure 45. *Forward plank.*

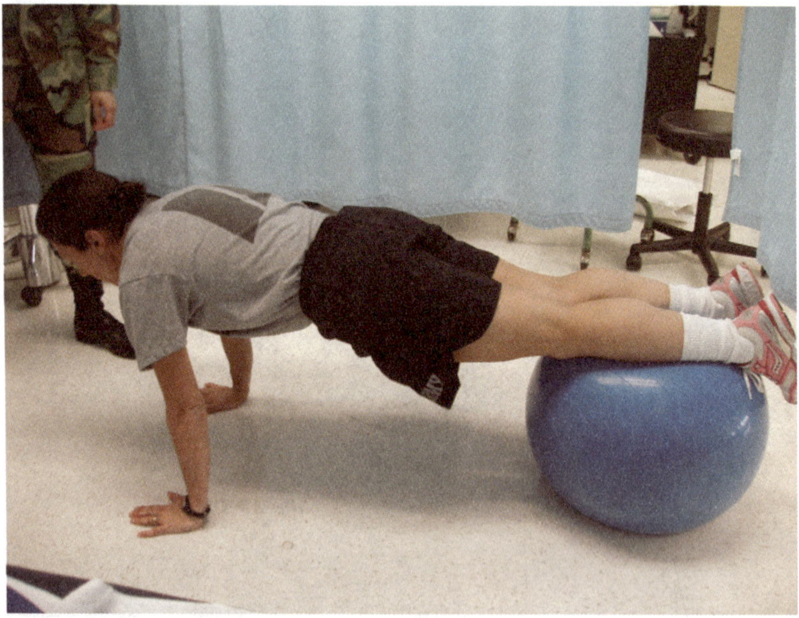

Figure 46. *Plank on ball.*

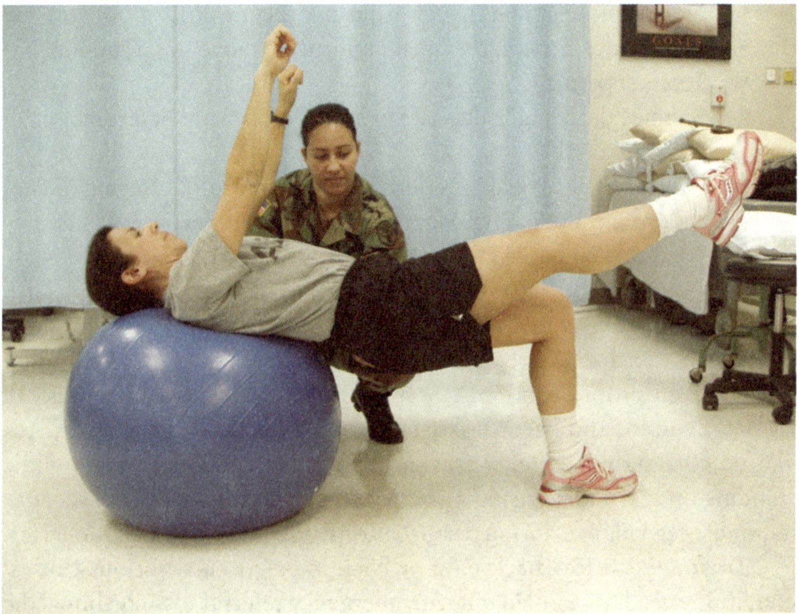

Figure 47. *Bridging on ball.*

Figure 48. *Prone alternating on ball with weights.*

- Side plank (Figure 44), forward plank (Figure 45).
- Having the physical therapist perform rhythmic stabilization.
- Prone superman.
- Supine "dead bug": alternate hands to knees without resting feet all the way to table.
- Advanced physioball positions: plank (Figure 46), bridging (Figure 47), prone alternating (Figure 48), and use of heavier weights.
- Medicine ball drills.

The next step is to incorporate core stabilization into all agility, sports-specific drills, strength training, plyometrics, power training, balance, and proprioception. Rotational drills with medicine balls are frequently used, and partner-assisted/group exercises are common. Advanced core-stabilizing exercises in this phase include balancing while standing on a dynadisc, using a physioball during single-leg wall sits, doing single-leg balance during perturbations, and performing plyometrics.

Treatment. If low back pain or back injury occurs, serious causes/conditions need to be ruled out. An emergency referral to an orthopaedic surgeon or neurosurgeon is necessary if there are complaints of bowel and/or bladder loss of function. Other reasons for referral include progressive numbness and/or weakness in one or both lower extremities, pain radiating past the knee that is not relieved with conservative treatment, or true night pain that causes arousal from a sound sleep. Additionally, severe or chronic back pain of unknown origin should be referred.[214]

Most acute low back pain is not serious and should be treated early with "active rest" (cross-training using pain-free forms of exercise, such as walking or swimming); ice; NSAIDs; and PT for evaluation, possible manipulation, education about exacerbation, reinjury prevention, ergonomics, body mechanics, stretching, and lumbar stabilization. Often, the injured military service member will require limited duty (physical profile) and/or day(s) off (quarters), especially if the training or job requirements are very physically demanding. However, the injured service member must be shown mobility, strengthening, and alternate cardiovascular exercises as soon as possible after injury or onset of pain. The primary goal is to prevent chronic low back pain and recurrences. Physical therapists can also evaluate and identify those who would benefit from spinal manipulation and specific strengthening, stretching, and range-of-motion exercises.

Spinal Manipulation. Spinal manipulation has been found to be very effective for a subgroup of patients suffering from low back pain. Flynn et al[156] developed a clinical prediction rule to help identify those with low back pain who would most likely benefit from manipulation. Spinal manipulation is clearly indicated if at least 4 of 5 of the following criteria are met: (1) symptom duration less than 16 days, (2) no symptoms distal to the knee, (3) a score of less than 19 on a fear-avoidance measure, (4) at least one hypomobile lumbar segment, and (5) at least one hip with more than 35 degrees of internal rotation.[156,215] The presence of at least 4 of 5 of these variables increased the likelihood of success, with spinal manipulation from 45% to 95%. The specific manipulation technique described in these studies has the patient positioned supine and passively side-bent away from the therapist. The therapist stands on the opposite side to be manipulated. The therapist then rotates the patient's upper body toward the therapist and then performs a quick posterior thrust through the anterior superior iliac spine.

Core Stabilization Exercises. Core stabilization exercises, as described previously, have been advocated in the treatment of those with low back pain.[216–219] The exercise regimen of strengthening the TrA and MF is partially based on the emerging research related to the role of these muscles in spinal stability. In addition, researchers have found that contraction of the TrA during the abdominal drawing-in maneuver instead of a more global bracing contraction of the global abdominal muscles provided greater SIJ stability.[220]

Recent researchers reported that lumbar stabilization programs resulted in better outcomes, compared with generalized exercise programs.[221,222] The theory behind stability exercises is that improved neuromuscular function will enhance or restore intrinsic support or motor control of the spine and pelvis, decrease the involuntary or subtle shear forces and irritation, and ultimately lead to pain relief.[221] A rational approach to safely enhancing lumbar stability is through endurance versus strictly strength training of the local deep core trunk muscles, which act to ensure neutral spine posture when under constant load.[223]

As described previously, under normal conditions, the TrA activates in a protective manner in anticipation of trunk and limb movement. However, this pattern of movement is not present in individuals with low back pain. In particular, in individuals with low back pain, the anticipatory contraction of the TrA is absent, and it has been suggested that the lack of this protective phenomenon may result in the recurrent acute episodes of low back pain commonly seen in the general population.[224–227] The additional primary trunk muscles (the external oblique, the internal oblique, the rectus abdominis , and the MF) were unable to be classified as anticipatory for all limb and trunk movements and, therefore, strengthens the construct that the TrA muscle is a key muscle in the maintenance of lumbar stability during dynamic tasks.

Posteriorly, the MF muscle is the main focus in rehabilitation of those with low back pain. It is thought to be specifically important for stabil-

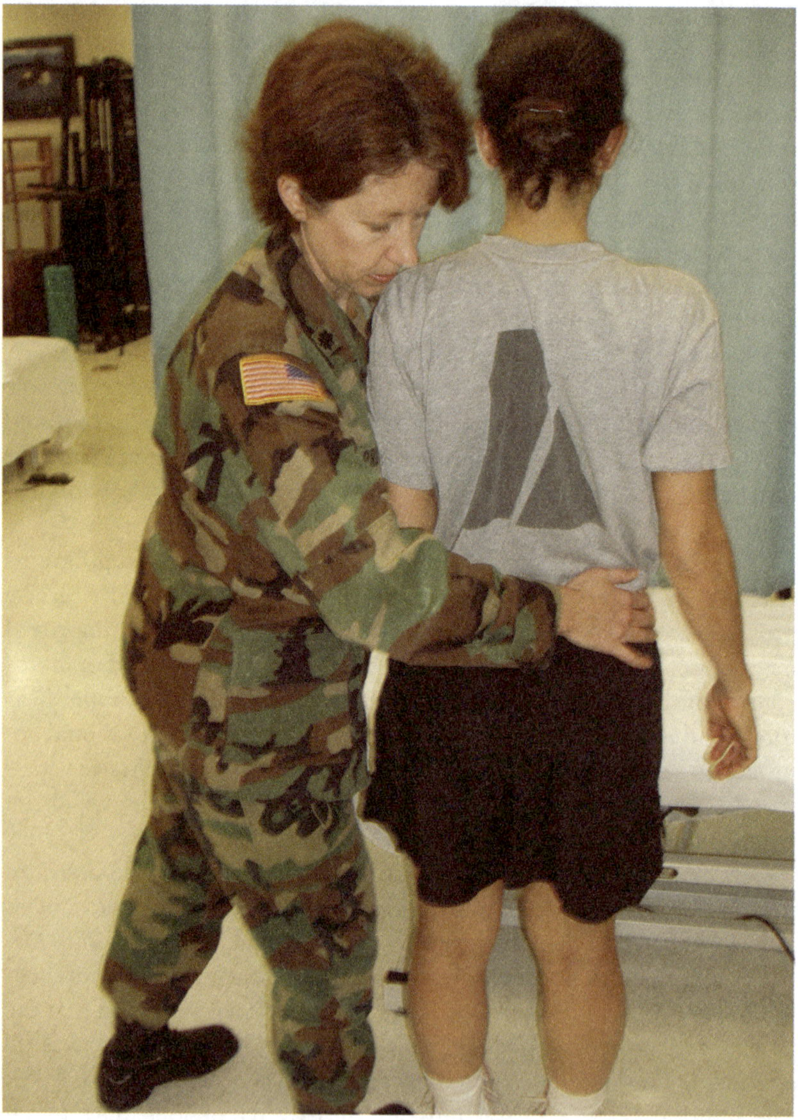

Figure 49. *Right lateral shift correction—start.*

ity, and it is the largest and most medial of the lumbar back muscles.[228] MF's function is to reduce anterior shear force of a functional spinal unit, primary trunk extensor, and a trunk stabilizer. Hides et al[229] reported that individuals without a history of low back pain had an average side-to-side difference of 3% ± 4% in the cross-sectional area of the MF

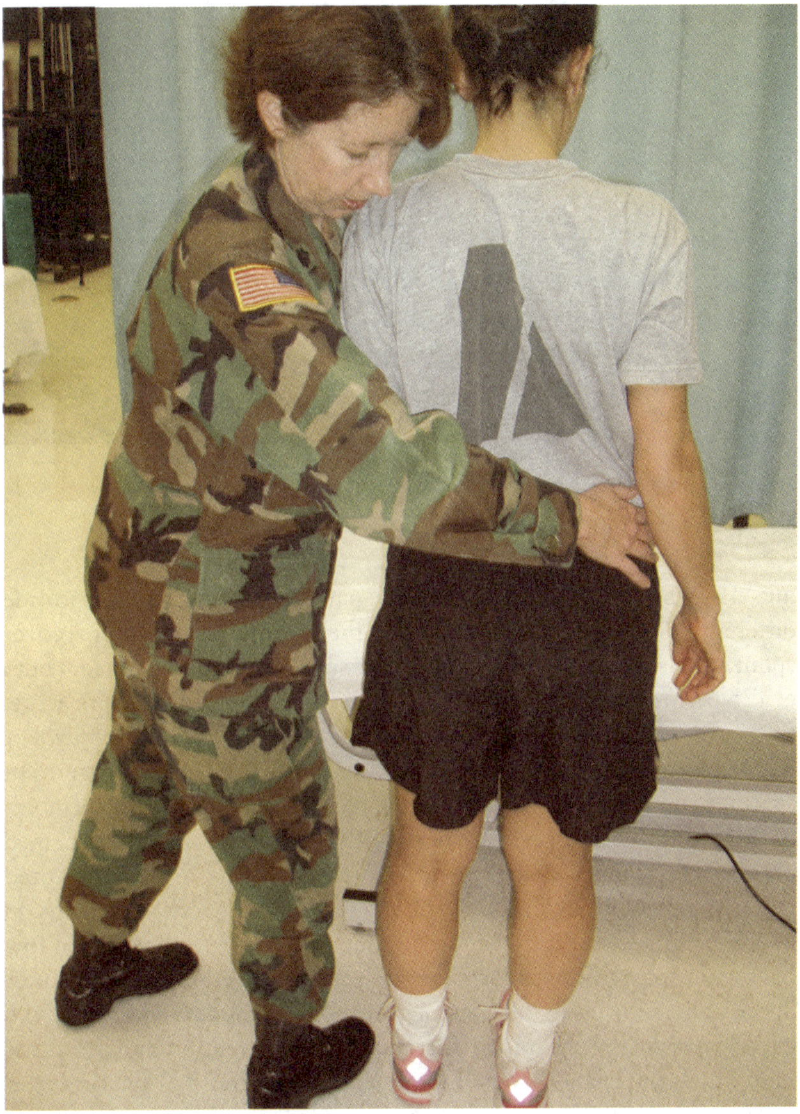

Figure 50. *Right lateral shift correction—end.*

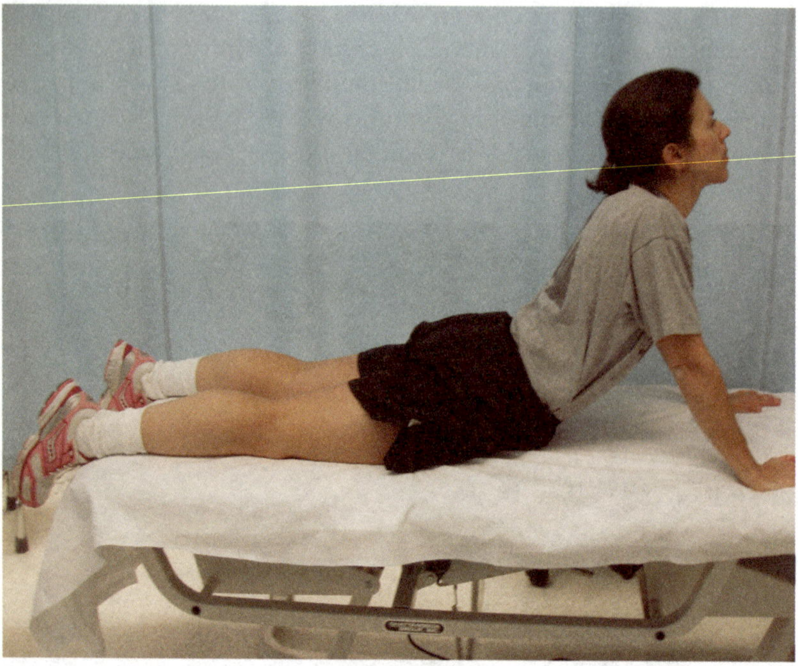

Figure 51. *Passive prone extension.*

muscles, whereas those with low back pain had an average side-to-side difference of 31% ± 8%. After injury, the MF muscle does not recover spontaneously with time and needs to be specifically strengthened. Hides et al[230] performed a randomized clinical trial with 39 subjects with acute, first-episode unilateral low back pain randomly assigned to a control or treatment group. The treatment group performed a 4-week program of specific localized MF exercises, and low back pain subsided in all members of this group. The reduced size of the MF in the control group remained practically unchanged, whereas the MF cross-sectional area was restored in the treatment group and was evident at 4- and 10-week follow-ups. In addition, subjects in the treatment group had fewer recurrences of low back pain than the control group subjects over the longitudinal study. Specifically, subjects in the control group were 12.4 times more likely to experience recurrences in the first year postintervention and 5.9 times more likely at 2- and 3-year follow-ups. Therefore, it is important to perform core stabilization exercises to not only help recover from low back conditions, but also to prevent recurrences.

Other Exercises. It is beyond the scope of this monograph to discuss all treatments for low back pain. However, flexibility and positional exercises are often prescribed for treatment of those with low back pain. For example, McKenzie lateral shift and/or extension exercises are often prescribed for those patients with intervertebral disc syndrome who get relief from positional change or directional movements, such as extension of the lumbar spine.[231] For those patients presenting with an acute lateral shift or list, the underlying cause is thought to be related to a posterior-lateral shifting of the nucleus pulposus within the disc, which puts pressure on pain-sensitive structures, such as nerve roots.[231] McKenzie and May[231] advocate manual or self-correction of the lateral shift followed by passive extension exercises that are thought to move the nucleus of the disc centrally (Figures 49 to 51). This procedure may increase low back pain initially, but should not increase leg pain. It is considered successful if the leg pain is relieved or eliminated, and lumbar motion is restored. Flexibility and positional exercises should be taught in conjunction with strengthening exercises.

Summary

Musculoskeletal injuries in military women are common. Prevention and management of such injuries are very important to sustain the fighting force and maintain military readiness. This monograph provides information about the incidence, risk factors, prevention, diagnosis, evaluation, treatment, and rehabilitation of common musculoskeletal overuse and traumatic injuries sustained by women in the military.

REFERENCES

1. US Department of the Army. *Prevention and Control of Musculoskeletal Injuries Associated with Physical Training*. Washington, DC: DA; 2006. Technical Bulletin Medical 592 (TB MED 592).
2. Bensel C, Kish R. *Lower Extremity Disorders Among Men and Women in Army Basic Training and Effects of Two Types of Boots*. Natick, MA: US Army Research Institute of Environmental Medicine; 1983. Technical Report TR-83/026.
3. Gardner L, Dziados J, Jones B, et al. Prevention of lower extremity stress fractures: a controlled trial of a shock absorbent insole. *Am J Public Health*. 1988;78:1563–1567.
4. Jones B, Bovee M, Harris J, Cowan D. Intrinsic risk factors for exercise-related injuries among male and female Army trainees. *Am J Sports Med*. 1993;21:705–710.
5. Jones B, Cowan D, Tomlinson J, Robinson J, Polly D, Frykman P. Epidemiology of injuries associated with physical training among young men in the Army. *Med Sci Sports Exerc*. 1993;25:197–203.
6. Knapik J, Cuthie J, Canham-Chervak M, et al. *Injury Incidence, Injury Risk Factors, and Physical Fitness of U.S. Army Basic Trainees at Ft. Jackson, SC, 1997*. Aberdeen Proving Ground, MD: US Army Center for Health Promotion and Preventive Medicine; 1998. Epidemiological Consultation Report 29-HE-7513-98.
7. Knapik J, Darakjy S, Scott S, et al. *Evaluation of Two Army Fitness Programs: The TRADOC Standardized Physical Training Program for Basic Combat Training and the Fitness Assessment Program*. Aberdeen Proving Ground, MD: US Army Center for Health Promotion and Preventive Medicine; 2004. Technical Report 12-HF-5772B-04.

8. Knapik J, Hauret K, Arnold S, et al. Injury and fitness outcomes during implementation of Physical Readiness Training. *Int J Sports Med.* 2003;24:372–381.
9. Knapik J, Sharp M, Canham-Chervak M, Hauret K, Patton J, Jones B. Risk factors for training-related injuries among men and women in Basic Combat Training. *Med Sci Sports Exerc.* 2001;33:946–954.
10. Kowal D. Nature and causes of injuries in women resulting from an endurance training program. *Am J Sports Med.* 1980;8:265–269.
11. Henderson N, Knapik J, Shaffer S, McKenzie T, Schneider G. Injuries and injury risk factors among men and women in U.S. Army Combat Medic Advanced Individual Training. *Mil Med.* 2000;165:647–652.
12. Bell N, Mangione T, Hemenway D, Amoroso P, Jones B. *High Injury Rates Among Female Army Trainees: A Function of Gender?* Natick, MA: US Army Research Institute of Environmental Medicine; 1996. Technical Report MISC96-6.
13. Knapik J, Canham-Chervak M, Hauret K, Hoedebecke E, Laurin M, Cuthie J. Discharges during U.S. Army Basic Training: injury rates and risk factors. *Mil Med.* 2001;166:641–647.
14. Summer G. Gender differences in the utilization of a military ship's medical department. *Mil Med.* 2001;166:32–33.
15. Hauret K, Shippey D, Knapik J. The Physical Training and Rehabilitation Program: duration of rehabilitation and final outcome of injuries in Basic Combat Training. *Mil Med.* 2001;166:820–826.
16. Knapik J, Sharp M, Canham M, et al. *Injury Incidence and Injury Risk Factors Among US Army Basic Trainees at Ft. Jackson, SC (Including Fitness Training Unit Personnel, Discharges, and Newstarts).* Aberdeen Proving Ground, MD: US Army Center for Health Promotion and Preventive Medicine; 1999. Epidemiological Consultation Report 29-HE-8370-99.
17. Acurater A. Biomechanics and the female athlete. In: Puhl J, Brown C, Vox R, eds. *Sports Perspectives for Women.* Champaign, IL: Human Kinetics; 1988.
18. Wilmore J. The application of science to sport: physiological profiles of male and female athletes. *Can J Appl Sport Sci.* 1979;4:103–115.
19. Jones B, Cowan D, Knapik J. Exercise, training and injuries. *Sports Med.* 1994;18:202–214.
20. Knapik J, Jones S, Darakjy S, et al. Injury rates and injury risk factors among United States Army wheel vehicle mechanics. *Mil Med.* 2007;172:988–996.
21. Knapik JJ, Hauret KG, Jones BH. Primary prevention of injuries in Initial Entry Training. In: DeKoning B, ed. *Recruit Medicine (Textbooks of Military Medicine).* Washington, DC: Department of the Army, Office of The Surgeon General, Borden Institute; 2006.
22. Delee J, Drez D, Miller M. *DeLee and Drez's Orthopaedic Sports Medicine.* Philadelphia, PA: WB Saunders; 2003.

23. Jones B, Harris J, Vinh T, Rubin C. Exercise-induced stress fractures and stress reactions of bone: epidemiology, etiology, and classification. In: Pandolf K, ed. *Exercise and Sport Sciences Review*. Baltimore, MD: Williams & Wilkins; 1989.
24. Davenport T, Kulig K, Matharu Y, Blanco C. The EdUReP model for nonsurgical management of tendinopathy. *Phys Ther.* 2005;85:1093–1103.
25. Brukner P, Bennell K. Stress fractures in female athletes: diagnosis, management and rehabilitation. *Sports Med.* 1997;24:419–429.
26. Martin A, McCulloch R. Bone dynamics: stress, strain and fracture. *J Sports Sci.* 1987;5:155–163.
27. Armstrong DW, Rue JP, Wilckens JH, Frassica FJ. Stress fracture injury in young military men and women. *Bone.* 2004;35:806–816.
28. Breithaupt J. Zur pathologie des menschlichen fuses. *Med Ztg Berlin.* 1855;24:169–171, 175–177.
29. Bennell K, Malcom S, Thomas S, Wark J, Brukner P. The incidence and distribution of stress fractures in competitive track and field athletes. *Am J Sports Med.* 1996;24:211–217.
30. Brunet M, Cook S, Brinker M, Dickinson J. A survey of running injuries in 1505 competitive and recreational runners. *J Sports Med Phys Fitness.* 1990;30:307–315.
31. Cameron K, Telford R, Wark J. Stress fractures in Australian competitive runners [abstract]. Paper presented at: Australian Sports Medicine Federation, Annual Scientific Conference in Sports Medicine; Perth, Australia; 1992.
32. Goldberg B, Pecora C. Stress fractures: a risk of increased training in freshmen. *Physician Sports Med.* 1994;22:68–78.
33. Johnson A, Weiss C, Wheeler D. Stress fractures of the femoral shaft in athletes—more common than expected: a new clinical test. *Am J Sports Med.* 1994;22:248–256.
34. Zernicke R, McNitt-Gray J, Otis C, et al. Stress fracture risk assessment among elite collegiate women runners. *Int Soc Biomech.* 1993;1506–1507.
35. Beck T, Ruff C, Shaffer R, Betsinger K, Trone D, Brodine S. Stress fracture in military recruits: gender differences in muscle and bone susceptibility factors. *Bone.* 2000;27:437–444.
36. Brudvig T, Gudger T, Obermeyer L. Stress fractures in 295 trainees: a one-year study of incidence as related to age, sex, and race. *Mil Med.* 1983;148:666–667.
37. Jones B, Thacker S, Gilchrist J, Kimsey C, Sosin D. Prevention of lower extremity stress fractures in athletes and soldiers: a systematic review. *Epidemiol Rev.* 2002;24:228–247.
38. Pester S, Smith P. Stress fractures in the lower extremities of soldiers in basic training. *Orthop Rev.* 1992;21:297–303.
39. Protzman R, Griffis C. Stress fractures in men and women undergoing military training. *J Bone Joint Surg Am.* 1977;59A:825–826.

40. Reinker KA, Ozburne S. A comparison of male and female orthopaedic pathology in basic training. *Mil Med.* 1979;144:532–536.
41. Bennell K, Malcolm S, Brukner P, et al. A 12-month prospective study of the relationship between stress fractures and bone turnover in athletes. *Calcif Tissue Int.* 1998;63:80–85.
42. Bennell K, Matheson G, Meeuwisse W, Brukner P. Risk factors for stress fractures. *Sports Med.* 1999;28:91–122.
43. Martin R. Determinants of the mechanical properties of bones. *J Biomech.* 1991;24(suppl 1):79–88.
44. Scully T, Besterman G. Stress fracture—a preventable training injury. *Mil Med.* 1982;147:285–287.
45. Cobb K, Bachrach L, Greendale G, et al. Disordered eating, menstrual irregularity, and bone mineral density in female runners. *Med Sci Sports Exerc.* 2003;35:711–719.
46. Winfield A, Moore J, Bracker M, Johnson C. Risk factors associated with stress reactions in female Marines. *Mil Med.* 1997;162:698–702.
47. Cline A, Jansen G, Melby C. Stress fractures in female Army recruits: implications of bone density, calcium intake, and exercise. *J Am Coll Nutr.* 1998;17:128–135.
48. Lappe J, Stegman M, Recker R. The impact of lifestyle factors on stress fractures in female Army recruits. *Osteoporos Int.* 2001;12:35–42.
49. Kaufman K, Brodine S, Shaffer R. Military training-related injuries surveillance, research, and prevention. *Am J Prev Med.* 2000;18(suppl 3):54–63.
50. Schwellnus M, Jordaan G, Noakes T. Prevention of common overuse injuries by the use of shock absorbing insoles. A prospective study. *Am J Sports Med.* 1990;18:636–641.
51. Shaffer R, Rauh M, Brodine S, Trone D, Macera C. Predictors of stress fracture susceptibility in young female recruits. *Am J Sports Med.* 2006;34:108–115.
52. Bennell K, Brukner P. Epidemiology and site specificity of stress fractures. *Clin Sports Med.* 1997;16:179–196.
53. Hulkko A, Orava S. Stress fractures in athletes. *Int J Sports Med.* 1987;8:221–226.
54. Noakes T, Smith J, Lindenberg G, Wills C. Pelvic stress fractures in long distance runners. *Am J Sports Med.* 1985;13:120–123.
55. O'Brien T, Wilcox N, Kersch T. Refractory pelvic stress fracture in a female long-distance runner. *Am J Orthop.* 1995;24:710–713.
56. Boden B, Osbahr D. High-risk stress fractures: evaluation and treatment. *J Am Acad Orthop Surg.* 2000;8:344–353.
57. Stanitski C, McMaster J, Scranton P. On the nature of stress fractures. *Am J Sports Med.* 1978;6:391–396.
58. Barrow G, Saha S. Menstrual irregularity and stress fractures in collegiate female distance runners. *Am J Sports Med.* 1988;16:209–216.

59. Prather J, Nusynowitz M, Snowdy H, Hughes A, McCartney W, Bagg R. Scintigraphic findings in stress fractures. *J Bone Joint Surg Am.* 1977;59:869–874.
60. Roub L, Gumerman L, Hanley E, Clark N, Goodman M, Herbert D. Bone stress: a radionuclide image perspective. *Radiology.* 1979;132:431–438.
61. Rupani H, Holder L, Espinola D, Engin S. Three-phase radionuclide bone imaging in sports medicine. *Radiology.* 1985;156:187–196.
62. Sterling J, Edelstein D, Calvo R, Webb R. Stress fractures in the athlete: diagnosis and management. *Sports Med.* 1992;14:336–346.
63. Fredericson M, Bergman A, Hoffman K, Dillingham M. Tibial stress reaction in runners: correlation of clinical symptoms and scintigraphy with a new magnetic resonance imaging grading system. *Am J Sport Med.* 1995;23:472–481.
64. Lee J, Yao L. Stress fractures: MR imaging. *Radiology.* 1988;169:217–220.
65. Tyrrell P, Davies A. Magnetic resonance imaging appearances of fatigue fractures of the long bones of the lower limb. *Br J Radiol.* 1994;67:332–338.
66. Kelly E, Jonson S, Cohen M, Shaffer R. Stress fractures of the pelvis in female Navy recruits: an analysis of possible mechanisms of injury. *Mil Med.* 2000;165:142–147.
67. Leveton A. March fractures of the long bones of the lower extremity and pelvis. *Am J Surg.* 1946;71:222–232.
68. Nickerson H. March fracture or insufficiency fracture. *Am J Surg.* 1943;62:154–164.
69. Selakovich W, Love L. Stress fractures of the pubic ramus. *J Bone J Surg Am.* 1954;36:573–576.
70. Wachsmuth W. Zur atiologic der schleichenden frakturen. *Der Chirung.* 1937;9:16–24.
71. Latshaw R, Kantner T, Kalenak A, Baum S, Corcoran J. A pelvic stress fracture in a female jogger. A case report. *Am J Sports Med.* 1981;9:54–56.
72. Orava S, Puranen J, Ala-Ketola L. Stress fractures caused by physical exercise. *Acta Orthop Scand.* 1978;49:19–27.
73. Pavlov H, Nelson T, Warren R, Torg J, Burstein A. Stress fractures of the pubic ramus. A report of twelve cases. *J Bone Joint Surg Am.* 1982;64:1020–1025.
74. Thorne D, Datz F. Pelvic stress fracture in female runners. *Clin Nucl Med.* 1986;11:828-829.
75. Forcher-Mayr O. Beitrag zu den trainingveranderungen beim sportfechten. *Wien Klin Wochenschr.* 1954;63:331–333.
76. Teitz C, Hu S, Arendt E. The female athlete: evaluation and treatment of sports-related problems. *J Am Acad Orthop Surg.* 1997;5:87–96.
77. Hill P, Chatterji S, Chambers D. Stress fracture of the pubic ramus in female recruits *J Bone Joint Surg Br.* 1996;78B:383–386.
78. Ozburn M, Nichols J. Pubic ramus and adductor insertion stress fractures in female basic trainees. *Mil Med.* 1981;146:332–334.
79. Shin A, Gillingham B. Fatigue fractures of the femoral neck in athletes. *J Am Acad Orthop Surg.* 1997;5:293–302.

80. Hajek M, Noble H. Stress fractures of the femoral neck in joggers: case reports and review of the literature. *Am J Sports Med.* 1982;10:112–116.
81. Anderson E. Fatigue fractures of the foot [abstract]. *Injury.* 1990;21:275–279.
82. Stienstra J, Gumann G. Phalangeal and metatarsal fractures. In: Guman G, ed. *Fractures of the Foot and Ankle.* Philadelphia, PA: Elsevier; 2004.
83. Slowik A. Stress fractures of the first metatarsal. *J Am Podiatry Assoc.* 1969;59:333–335.
84. Torg J, Balduini F, Zelko R, Pavlov H, Peff T, Das M. Fractures of the base of the fifth metatarsal distal to the tuberosity: classification and guidelines for non-surgical and surgical management. *J Bone Joint Surg Am.* 1984;66:209–214.
85. Thacker S, Gilchrist J, Stroup D, Kimsey D. The prevention of shin splints in sports: a systematic review of the literature. *Med Sci Sports Exerc.* 2002;34:32–40.
86. Turnipseed W. Clinical review of patients treated for atypical claudication: a 28-year experience. *J Vasc Surg.* 2004;40:79–85.
87. Houglum P, Porter D. Foot, ankle and leg injuries. In: Starkey C, Johnson G, eds. *Athletic Training and Sports Medicine.* Sudbury, MA: Jones and Bartlett; 2006.
88. van den Brand J, Nelson T, Verleisdonk E, van der werken C. The diagnostic value of intracompartmental pressure measurement, magnetic resonance imaging, and near-infrared spectroscopy in chronic exertional compartment syndrome: a prospective study in 50 patients. *Am J Sports Med.* 2005;33:699–704.
89. Mubarak S. Exertional compartment syndrome. In: D'Ambrosia RD, Drez D, eds. *Prevention and Treatment of Running Injuries.* Thorofare, NJ: Slack, Inc; 1989.
90. Styf J. Chronic exercise-induces pain in the anterior aspect of the lower leg. An overview of the diagnosis. *Sports Med.* 1989;7:331–339.
91. Pedowitz R, Hargens A, Mubarak S, Gershuni D. Modified criteria for the objective diagnosis of chronic syndrome of the leg. *Am J Sports Med.* 1990;18:35–40.
92. Rorabeck C, Bourne R, Fowler P, Finlay B, Nott L. The role of tissue pressure measurement in diagnosing chronic anterior compartment syndrome. *Am J Sports Med.* 1988;16:143–146.
93. Vath S, DeBerardino T, Stoneman P. *Prospective Analysis of Anterior and Lateral Leg Pain.* West Point, NY: The Department of Physical Therapy, Keller Army Community Hospital; 2005.
94. Messier S, Pittala K. Etiologic factors associated with selected running injuries. *Med Sci Sports Exerc.* 1988;20:501–505.
95. Andrish J, Bergfeld J, Walheim J. A prospective study on the management of shin splints. *J Bone Joint Surg Am.* 1974;56:1697–1700.
96. Rauh M, Margherita A, Rice S, et al. High school cross country running injuries: a longitudinal study. *J Clin Sports Med.* 2000;10:110–116.
97. Michael R, Holder L. The soleus syndrome. A cause of medial tibial stress (shin splints). *Am J Sports Med.* 1985;13:87–94.

98. McCarty E, Spindler K, Bartz R. Knee and leg: soft-tissue trauma. In: Vaccaro A, ed. *Orthopaedic Knowledge Update*. Rosemont, IL: The American Academy of Orthopaedic Surgeons; 2005.
99. Rubin C. Sports injuries in the female athlete. *N J Med.* 1991;88:643–645.
100. Post W. Anterior knee pain: diagnosis and treatment. *J Am Acad Orthop Surg.* 2005;13:534–543.
101. Kirk K, Kuklo T, Klemme W. Iliotibial band friction syndrome. *Orthopaedics.* 2000;23:1209–1217.
102. Colson J, Armour W. *Sports Injuries and Their Treatment*. Philadelphia, PA: JB Lippincott; 1961.
103. Holmes J, Pruitt A, Whalen N. Iliotibial band syndrome in cyclists. *Am J Sports Med.* 1993;21:419–421.
104. Orava S. Iliotibial band friction syndrome in athletes: an uncommon exertion syndrome on the lateral side of the knee. *Br J Sports Med.* 1978;12:69–73.
105. Renne J. The iliotibial band friction syndrome. *J Bone Joint Surg Am.* 1975;57:1110–1111.
106. Sutker A, Barber F, Jackson D, Pagliano J. Iliotibial band syndrome in distance runners. *Sports Med.* 1985;2:447–451.
107. Linenger J, Christensen C. Is iliotibial band syndrome overlooked? *Physician Sports Med.* 1992;20:98–108.
108. Grood E, Noyes F, Butler D, Suntay W. Ligamentous and capsular restraints preventing straight medial and lateral laxity in intact human cadaver knees. *J Bone Joint Surg Am.* 1981;63:1257–1269.
109. Terry G, Hughston J, Norwood L. The anatomy of the iliopatellar band and iliotibial tract. *Am J Sports Med.* 1986;14:39–45.
110. Ekman E, Pope T, Martin D, Curl W. Magnetic resonance imaging of iliotibial band syndrome. *Am J Sports Med.* 1994;22:851–854.
111. Murphy B, Hechtman K, Uribe J, Selesnick H, Smith R, Zlatkin M. Iliotibial band friction syndrome: MR imaging findings. *Radiology.* 1992;185:569–571.
112. Nishimura G, Yamato M, Tamai K, Takahashi J, Uetani M. MR findings in iliotibial band syndrome. *Skeletal Radiol.* 1997;26:533–537.
113. Nemeth W. Arthroscopic treatment of resistant iliotibial band friction syndrome [abstract]. *Orthop Trans.* 1992;16:46.
114. Jones D, James S. Overuse injuries of the lower extremity: shin splints, iliotibial band friction syndrome, and exertional compartment syndromes. *Clin Sports Med.* 1987;6:273–290.
115. Krivickas L. Anatomical factors associated with overuse sports injuries. *Sports Med.* 1997;24:132–146.
116. Messier S, Edwards D, Martin D, et al. Etiology of iliotibial band friction syndrome in distance runners. *Med Sci Sports Exerc.* 1995;27:951–960.
117. Noble C. Iliotibial band friction syndrome in runners. *Am J Sports Med.* 1980;8:232–234.

118. Clement D, Taunton J, Smart G, McNicol K. A survey of overuse running injuries. *Physician Sports Med.* 1981;9:47–58.
119. Safran M, Fu F. Uncommon causes of knee pain in the athlete. *Orthop Clin North Am.* 1995;26:547–559.
120. Ober F. The role of the iliotibial band and fascia lata as a factor in the causation of low back disabilities and sciatica. *J Bone Joint Surg Am.* 1936;18:105–110.
121. De Geeter F, De Neve J, van Steelandt H. Bone scan in iliotibial band friction syndrome. *Clin Nucl Med.* 1995;20:550–551.
122. Martens M, Wouters P, Burssens A, Muller J. Patellar tendinitis: pathology and results of treatment. *Acta Orthop Scand.* 1982;53:445–450.
123. Soslowsky L, Thomopoulos S, Tun S, et al. Neer Award 1999. Overuse activity injures the supraspinatus tendon in an animal model: a histologic and biomechanical study. *J Shoulder Elbow Surg.* 2000;9:79–84.
124. Hicks J. The mechanics of the foot. II, the plantar aponeurosis and the arch. *J Anat.* 1954;88:25–30.
125. Tomlinson J, Lednar W, Jackson J. Risk of injury in soldiers. *Mil Med.* 1987;152:60–64.
126. Beynnon B. Risk factors for knee ligament trauma. *J Orthop Sports Phys Ther.* 2003;33:A10–A13.
127. Fleming B. Biomechanics of the anterior cruciate ligament. *J Orthop Sports Phys Ther.* 2003;33:A13–A15.
128. Davis I, Ireland M. ACL injuries—the gender bias. Proceedings and abstracts of Research Retreat. II, April 4–5, 2003. Lexington, KY. *J Orthop Sports Phys Ther.* 2003;33:A1–A30.
129. Almeida S, Trone D, Leone D, Shaffer R, Patheal S, Long K. Gender differences in musculoskeletal injury rates: a function of symptom reporting? *Med Sci Sports Exerc.* 1999;31:1807–1812.
130. Jackson D, Ashley R, Powell J. Ankle sprains in young athletes: relation of severity and disability. *Clin Orthop Relat Res.* 1974;101:201–215.
131. Gerber J, Williams G, Scoville C, Arciero R, Taylor D. Persistent disability associated with ankle sprains: a prospective examination of an athletic population. *Foot Ankle Int.* 1998;19:653–660.
132. Amoroso P, Bell N, Jones B. Injury among female and male parachutists. *Aviat Space Environ Med.* 1997;68:1006–1011.
133. Craig S, Morgan J. Parachuting injury surveillance, Fort Bragg, North Carolina, May 1993 to December 1994. *Mil Med.* 1997;162:162–164.
134. Hallel T, Naggan L. Parachute injuries: a retrospective study of 83,718 jumps. *J Trauma.* 1975;15:14–19.
135. Kirby N. Parachuting injuries. *Proc R Soc Med.* 1974;67:17–21.
136. Hosea T, Carey C, Harrer M. The gender issue: epidemiology of ankle injuries in athletes who participate in basketball. *Clin Orthop Relat Res.* 2000;372:45–49.
137. Fitzgerald RJ. Acetabular labrum tears: diagnosis and treatment. *Clin Orthop Relat Res.* 1995;311:60–68.

138. Hase T, Ueo T. Acetabular labral tear: arthroscopic diagnosis and treatment *Arthroscopy.* 1999;15:138–141.
139. McCarthy J, Noble P, Schuck M, Wright J, Lee J. The Otto E. Aufranc Award: the role of labral lesions to development of early degenerative hip disease. *Clin Orthop Relat Res.* 2001;393:25–37.
140. Santori N, Villar R. Acetabular labral tears: result of arthroscopic partial limbectomy. *Arthroscopy.* 2000;16:11–15.
141. Lewis C, Sahrmann S. Acetabular labral tears. *Phys Ther.* 2006;86:110–121.
142. Narvani A, Tsiridis E, Kendall S, Chaudhuri R, Thomas P. A preliminary report on prevalence of acetabular labrum tears in sports patients with groin pain. *Knee Surg Sports Traumatol Arthrosc.* 2003;11:403–408.
143. Ross A, Javernick M, Freedman B, Springer B, Wieand K, Murphy K. Repair of the adult acetabular labrum: short-term outcome. *Am J Sports Med.* Forthcoming.
144. Feuerstein M, Berkowitz S, Peck C. Musculoskeletal-related disability in US Army personnel: prevalence, gender, and military occupational specialties. *J Occup Environ Med.* 1997;39:68–78.
145. Berkowitz S, Feuerstein M, Lopez M, Peck C. Occupational back disability in U.S. Army personnel. *Mil Med.* 1999;164:412–418.
146. Bernard T, Kirkaldy-Willis W. Recognizing specific characteristics of nonspecific low back pain. *Clin Orthop Relat Res.* 1987;217:266–280.
147. Cohen S. Sacroiliac joint pain: a comprehensive review of anatomy, diagnosis and treatment. *Anesth Analg.* 2005;1001:1440–1453.
148. Fortin J, Aprill C, Ponthieux B, Pier J. Sacroiliac joint: pain referral maps upon applying a new injection/arthrography technique. Part II, clinical evaluation. *Spine.* 1994;19:1483–1489.
149. Zelle B, Gruen G, Brown S, George S. Sacroiliac joint dysfunction. Evaluation and management. Review article. *Crit J Pain.* 2005;21:446–455.
150. Endresen E. Pelvic pain and low back pain in pregnant women: an epidemiological study. *Scand J Rheumatol.* 1995;24:135–141.
151. Kristiansson P, Svardsudd K, von Schoultz B. Back pain during pregnancy: a prospective study. *Spine.* 1996;21:702–709.
152. Larsen E, Wilken-Jensen C, Hansen A, et al. Symptom-giving pelvic girdle relaxation in pregnancy. I, prevalence and risk factors. *Acta Obstet Gynecol Scand.* 1999;78:105–110.
153. Ostgaard H, Andersson G, Karlsson K. Prevalence of back pain in pregnancy. *Spine.* 1991;16:549–552.
154. Bernard T, Cassidy J. The sacroiliac joint syndrome: pathophysiology, diagnosis, and management. In: Frymoyer J, ed. *The Adult Spine: Principles and Practice.* New York, NY: Raven Press Ltd; 1991: 2107–2130.
155. Dreyfuss P, Michaelsen M, Pauza K, McLarty J, Bogduk N. The value of medical history and physical examination in diagnosing sacroiliac joint pain. *Spine.* 1996;21:2594–2602.

156. Flynn T, Fritz J, Whitman J, et al. A clinical prediction rule for classifying patients with low back pain who demonstrate short-term improvement with spinal manipulation. *Spine*. 2002;27:2835-2843.
157. Laslett M, Williams M. The reliability of selected pain provocation tests for sacroiliac joint pathology. *Spine*. 1994;19:1243–1249.
158. Broadhurst N, Bond M. Pain provocative tests for the assessment of sacroiliac joint dysfunction. *J Spinal Disord*. 1998;11:341–345.
159. Ostgaard H, Zetherstrom G, Roos-Hansson E. The posterior pelvic pain provocation test in pregnant women. *Eur Spine J*. 1994;3:258–260.
160. Cibulka M, Koldenhoff R. Clinical usefulness of a cluster of sacroiliac joint tests in patients with and without low back pain. *J Orthop Sports Phys Ther.* 1999;29:83–92.
161. Knapik J, Darakjy S, Scott S, et al. Evaluation of a standardized physical training program for Basic Combat Training. *J Strength Cond Res*. 2005;19:246–253.
162. LaStayo P, Woolf J, Lewek M, Snyder-Mackler L, Reich T, Lindstedt S. Eccentric muscle contractions: their contribution to injury, prevention, rehabilitation, and sport. *J Orthop Sports Phys Ther.* 2003;33:557–570.
162a. Runner's World. Choosing a shoe: the very basics. http:www.runnersworld.co.uk/shoes/choosing-a-shoe-the-very-basics/481.html. Accessed June 4, 2010.
163. Rudzki S, Cummingham M. The effect of a modified physical training program in reducing injury and medical discharge rates in Australian Army recruits. *Mil Med*. 1999;164:648–652.
164. Knapik J, Bullock S, Canada S, Toney E, Wells J. *The Aberdeen Proving Ground Injury Control Project: Influence of a Multiple Intervention Program on Injuries and Fitness Among Ordnance School Students in Advanced Individual Training.* Aberdeen Proving Ground, MD: US Army Center for Health Promotion and Preventive Medicine; 2003. Report 12-HF-7990-03.
165. Knapik J, Bullock S, Canada S, et al. Influence of an injury reduction program on injury and fitness outcomes among soldiers. *Inj Prev.* 2004;10:37–42.
166. Knapik J, Hauret K, Bednarek J, et al. *The Victory Fitness Program. Influence of the US Army's Emerging Physical Fitness Doctrine on Fitness and Injuries in Basic Combat Training*. Aberdeen Proving Ground, MD: US Army Center for Health Promotion and Preventive Medicine; 2001. Epidemiological Consultation Report 12-MA-5762-01.
167. Shaffer R. Musculoskeletal Injury Project. Paper presented at: the 43rd Annual Meeting of the American College of Sports Medicine; Cincinnati, OH; 1996.
168. Thompson K, McVey E. Efficacy of shock absorbing boots to reduce lower extremity injuries in female Navy recruits [abstract 1780]. *Med Sci Sports Exerc.* 2004;36:S261.
169. Pope R. Prevention of pelvic stress fractures in female Army recruits. *Mil Med.* 1999;164:370–373.
170. Green N, Rogers R, Lipscomb A. Nonunions of stress fractures of the tibia. *Am J Sports Med*. 1985;13:171–176.

171. Orava S, Hulkko A. Stress fractures in athletes. *Int J Sports Med.* 1987;8:221–226.
172. Dameron T. Fractures of the proximal fifth metatarsal: selecting the best treatment option. *J Am Acad Orthop Surg.* 1995;3:110–114.
172a. Den Hartog BD. Fracture of the proximal fifth metatarsal. *J Am Acad Orthop Surg.* 2009;17:458–464.
173. Fronek J, Mubarak S, Hargens A, et al. Management of chronic exertional anterior compartment syndrome of the lower extremity. *Clin Orthop Relat Res.* 1987;220:217–227.
174. Martens M, Backaert M, Vermaut G, Mulier J. Chronic leg pain in athletes due to a recurrent compartment syndrome. *Am J Sports Med.* 1984;12:148–151.
175. Rorabeck C, Fowler P, Nott L. The results of fasciotomy in the management of chronic exertional compartment syndrome. *Am J Sports Med.* 1988;16:224–227.
176. McConnell J. The management of chondromalacia patellae: a long term solution. *Aust J Physiother.* 1986;32:215–223.
177. Dorotka R, Jimenez-Boj E, Kypta A, Kollar B. The patellofemoral pain syndrome in recruits undergoing military training: a prospective 2-year follow-up study. *Mil Med.* 2003;168:337–340.
178. Schwellnus M, Theunissen L, Noakes T, Reinach S. Anti-inflammatory and combined and anti-inflammatory/analgesic medication in early management of iliotibial band friction syndrome. A clinical trial. *South Afr Med J.* 1991;79:602–606.
179. Fredericson M, Cookingham C, Chaudhari A, Dowdell B, Oestreicher N, Sahrmann S. Hip abductor weakness in distance runners with iliotibial band syndrome. *Clin J Sport Med.* 2000;10:169–175.
180. Aronen J, Chronister R, Regan K, Hensien M. Practical, conservative management of iliotibial band syndrome. *Physician Sports Med.* 1993;21:59–69.
181. Martens M, Libbrecht P, Brussens A. Surgical treatment of the iliotibial band friction syndrome. *Am J Sports Med.* 1989;17:651–654.
182. Cook J, Purdam C. Rehabilitation of lower limb tendinopathies. *Clin J Sport Med.* 2003;22:777–789.
183. Bolgla L, Keskula D. Reliability of lower extremity functional performance tests. *J Orthop Sports Phys Ther.* 1997;26:138–142.
184. Curwin S, Stanish W. *Tendinitis: Its Etiology and Treatment.* Lexington, KY: Collamore Press; 1984.
185. Alfredson H. Chronic midportion Achilles tendinopathy: an update on research and treatment. *Clinics Sports Med.* 2003;22:727–741.
186. Alfredson H, Pietilä T, Jonsson P, Lorentzon R. Heavy-load eccentric calf muscle training for the treatment of chronic Achilles tendinosis. *Am J Sports Med.* 1998;26:360–366.
187. Cannell L, Taunton J, Clement D, et al. A randomised clinical trial of the efficacy of drop squats or leg extension/leg curl exercises to treat clinically diagnosed jumper's knee in athletes: pilot study. *Br J Sports Med.* 2001;35:60–64.

188. Purdam C, Jonsson P, Alfredson H, et al. A pilot study of the eccentric decline squat in the management of painful chronic patellar tendinopathy. *Br J Sports Med.* 2004;38:395–397.
189. Stasinopoulos D, Stasinopoulos I. Comparison of effects of exercise programme, pulsed ultrasound and transverse friction in the treatment of chronic patellar tendinopathy. *Clin Rehabil.* 2004;18:347–352.
190. Young M, Cook J, Purdam C, Kiss Z, Alfredson H. Eccentric decline squat protocol offers superior results at 12 months compared with traditional eccentric protocol for patellar tendinopathy in volleyball players. *Br J Sports Med.* 2005;39:102–105.
191. Hewett T, Stroupe A, Nance T, Noyes F. Plyometric training in female athletes: decreased impact forces and increased hamstring torques. *Am J Sports Med.* 1996;24:765–773.
192. Pope R. Rubber matting on an obstacle course causes anterior cruciate ligament ruptures and its removal eliminates them. *Mil Med.* 2002;167:355–358.
193. Amoroso P, Ryan J, Bickley B, Leitschuh P, Taylor D, Jones B. Braced for impact: reducing paratrooper's ankle sprains using outside-the-boot braces. *J Trauma.* 1998;45:575–580.
194. Schmidt M, Sulsky S, Amoroso P. Effectiveness of an outside-the-boot ankle brace in reducing parachute related ankle injuries. *Inj Prev.* 2005;11:163–168.
195. Sitler M, Ryan J, Wheeler B, et al. The efficacy of a semirigid ankle stabilizer to reduce acute ankle injury in baskeball players. *Am J Sports Med.* 1994;22:582–585.
196. Willems T, Witvrouw E, Delbaere K, Philippaerts R, De Bourdeaudhuij I, De Clercq D. Intrinsic risk factors for inversion ankle sprains in females—a prospective study. *Scand J Med Sci Sports.* 2005;15:336–345.
197. Springer BA, Arciero RA, Tenuta JJ, Taylor DC. A prospective study of modified Ottawa Ankle Rules in a military population. *Am J Sports Med.* 2000;28:864–888.
198. Mahieu N, Witvrouw E, Stevens V, Van Tiggelen D, Roget P. Intrinsic risk factors for the development of Achilles tendon overuse injury: a prospective study. *Am J Sports Med.* 2006;34:226–235.
199. Mason J. Acetabular labral tears in the athlete. *Clin Sports Med.* 2001;20:779–790.
200. Hides J, Jull G, Richardson C. Long-term effects of specific stabilizing exercises for first-episode low back pain. *Spine.* 2001;26:E243–E248.
201. Gundewall B, Liljeqvist M, Hansson T. Primary prevention of back symptoms and absence from work. A prospective randomized study among hospital employees. *Spine.* 1993;18:587–594.
202. Ebenbichler GR, Oddsson LI, Kollmitzer J, Erim Z. Sensory-motor control of the lower back: implications for rehabilitation. *Med Sci Sports Exerc.* 2001;33:1889–1898.
203. Garner-Morse M, Stokes I. The effects of abdominal muscle coactivation on lumbar spine stability. *Spine.* 1998;23:86–92.

204. Ebenbichler G, Oddsson L, Kollmitzer J, Erim Z. Sensory-motor control of the lower back: implications for rehabilitation. *Med Sci Sports Exerc.* 2001;33:1889–1898.
205. Panjabi M. The stabilizing system of the spine, Part 1: function, dysfunction, adaption, and enhancement. *J Spinal Disord.* 1992;5:383–389.
206. Hodges P, Richardson C. Contraction of the abdominal muscles associated with movement of the lower limb. *Phys Ther.* 1997;77:132–142.
207. Hodges P, Richardson C. Feedforward contraction of transversus abdominis is not influenced by the direction of arm movement. *Exp Brain Res.* 1997;114:362–370.
208. Hodges P, Richardson C. Inefficient muscular stabilisation of the lumbar spine associated with low back pain: a motor control evaluation of transversus abdominis. *Spine.* 1996;21:2640–2650.
209. Hodges P, Richardson C. Relationship between limb movement speed and associated contraction of the trunk muscles. *Ergonomics.* 1997;40:1220–1230.
210. Richardson C, Hodges P, Hides J. *Therapeutic Exercise for Lumbopelvic Stabilization. A Motor Control Approach for the Treatment and Prevention of Low Back Pain.* New York, NY: Churchill Livingstone: Elsevier; 2004.
211. Macintosh J, Valencia F, Bogduk N, Munro R. The morphology of the human lumbar multifidus. *Clin Biomech.* 1986;1:196–204.
212. Hodges P, Jull G. Motor relearning strategies for the rehabilitation of intervertebral control of the spine. In: Liebenson C, ed. *Rehabilitation of the Spine: A Practitioner's Manual.* Baltimore, MD: Lippincott Williams & Wilkins; 2003.
213. Teyhen D, Miltenberger C, Deiters H, et al. The use of ultrasound imaging of the abdominal drawing-in maneuver in subjects with low back pain. *J Orthop Sports Phys Ther.* 2005;35:346–355.
214. Bono C. Low back pain in athletes. *J Bone Joint Surg Am.* 2004;86:382–396.
215. Childs J, Fritz J, Flynn T, et al. A clinical prediction rule to identify patients with low back pain most likely to benefit from spinal manipulation: a validation study. *Ann Intern Med.* 2004;141:920–928.
216. Akuthota V, Nadler S. Core strengthening. *Arch Phys Med Rehabil.* 2004;85:S86–S92.
217. Hicks G, Fritz J, Delitto A, McGill S. Preliminary development of a clinical prediction rule for determining which patients will respond to a stabilization exercise program. *Arch Phys Med Rehabil.* 2005;86:1753–1762.
218. Richardson C, Jull J, Hodges P, Hides J. *Therapeutic Exercise for Spinal Segmental Stabilization in Low Back Pain: Scientific Basis and Clinical Approach.* Philadelphia, PA: Churchill Livingstone; 1999. Chap 11.
219. Saal J. Dynamic muscular stabilization in the nonoperative treatment of lumbar pain syndromes. *Orthop Rev.* 1990;19:691–700.
220. Richardson C, Snijders C, Hides J, Damen L, Pas M, Storm J. The relation between the transversus abdominis muscles, sacroiliac joint mechanics, and low back pain. *Spine.* 2002;27:399–405.

221. Hodges P. Core stability exercise in chronic low back pain. *Orthop Clin North Am.* 2003;34:245–254.
222. O'Sullivan P, Phyty G, Twomey L, Allison G. Evaluation of specific stabilizing exercise in the treatment of chronic low back pain with radiologic diagnosis of spondylolysis or spondylolisthesis. *Spine.* 1997;22:2959–2967.
223. McGill S. Low back stability: from formal description to issues for performance and rehabilitation. *Exerc Sport Sci Rev.* 2001;29:26–31.
224. Hodges P. The role of the motor system in spinal pain: implications for rehabilitation of the athlete following lower back pain. *J Sci Med Sport.* 2000;3:243–253.
225. Hodges P, Richardson C. Altered trunk muscle recruitment in people with low back pain with upper limb movement at different speeds. *Arch Phys Med Rehabil.* 1999;80:1005–1012.
226. Hodges P, Richardson C. Delayed postural contraction of transversus abdominis in low back pain associated with movement of the lower limb. *J Spinal Disord.* 1998;11:46–56.
227. Richardson C, Jull G, Hodges P, Hides J. *Therapeutic Exercise for Spinal Segmental Stabilization in Low Back Pain: Scientific Basis and Clinical Approach.* Philadelphia, PA: Churchill Livingstone; 1999. Chap 2.
228. Lewin T, Moffet B, Vidik A. The morphology of the lumbar synovial intervertebral joints. *Acta Morph Neerl Scand.* 1962;4:299–319.
229. Hides J, Stokes M, Saide M, Jull G, Cooper D. Evidence of lumbar multifidus muscle wasting ipsilateral to symptoms in patients with acute/subacute low back pain. *Spine.* 1994;19:165–172.
230. Hides J, Richardson C, Jull G. Multifidus muscle recovery is not automatic after resolution of acute, first-episode low back pain. *Spine.* 1996;21:2763–2769.
231. McKenzie R, May S. *The Lumbar Spine: Mechanical Diagnosis & Therapy.* Waikanae, New Zealand: Spinal Publications; 1981.

Appendix: Postoperative Protocol—Hip Arthroscopy for Acetabular Labral Lesions

PHYSICAL THERAPY SERVICE
WALTER REED ARMY MEDICAL CENTER
WASHINGTON, DC

Preoperative Visit

- Discuss role of Physical Therapy postoperatively
- Teach normal heel-toe gait with crutches on level surfaces and stairs
- Instruct in knee strengthening, hip range of motion, and stretching exercises (postop protocol)
- Discuss importance of postop portal site management to prevent adhesions/pain
- Schedule postop follow-up

Phase I (0–1 Week)

NOTE: Exercise prescription is dependent on the tissue healing process and *individual* functional readiness in *all* stages. If any concerns or complications arise regarding the progress of any patient, Physical Therapy will contact the orthopaedic surgeon.

- Weight-bearing as tolerated using crutches with normal heel-to-toe gait
- Isometric quadriceps/gluteus/hamstring/adductor/abductor sets
- Active assistive range of motion: heel slides with sheet, supine hip abduction and adduction
- Bridging
- Weight-shifting
- Single-leg balance
- Standing hip abduction/adduction, flexion, extension without resistance (stand on affected side: closed-chain perturbations)
- Grade I to grade II inferior/posterior mobilizations for pain
- Portal site (scope site) scar massage when closed

Phase II (2–3 Weeks)

- Progress to ambulation without crutches when normal gait is present
- Continue hip range of motion with gradual end-range stretch (aggressive stretch is not necessary)
- Single-leg bridging
- Stationary bike as tolerated
- Aquatic therapy
- Begin lumbar stabilization

Phase III (4–6 Weeks)

- Continue hip stretching
- Mini-squats
- Leg press
- Backward lunges in pain-free range
- Step-downs
- Closed-chain hip flexion, extension, abduction/adduction: use tubing/bands on unaffected limb for resistance
- StairMaster/cross-trainer/elliptical

Phase IV (6+ Weeks)

- Gradual progression to full functional activities as tolerated
- Begin walk-to-jog progression when pain-free

This Appendix was adapted with permission from the US Army Medical Department, Physical Therapy Service, Walter Reed Army Medical Center, Washington, DC.